Marco Aurélio Vianello
Elza Massae Mamizuka

Virulence factors of Staphylococcus aureus and its regulator agr

Marco Aurélio Vianello
Elza Massae Mamizuka

Virulence factors of Staphylococcus aureus and its regulator agr

Various clinical manifestations from the same pathogen

ScienciaScripts

Imprint
Any brand names and product names mentioned in this book are subject to trademark, brand or patent protection and are trademarks or registered trademarks of their respective holders. The use of brand names, product names, common names, trade names, product descriptions etc. even without a particular marking in this work is in no way to be construed to mean that such names may be regarded as unrestricted in respect of trademark and brand protection legislation and could thus be used by anyone.

Cover image: www.ingimage.com

This book is a translation from the original published under ISBN 978-3-330-77195-6.

Publisher:
Sciencia Scripts
is a trademark of
Dodo Books Indian Ocean Ltd. and OmniScriptum S.R.L publishing group

120 High Road, East Finchley, London, N2 9ED, United Kingdom
Str. Armeneasca 28/1, office 1, Chisinau MD-2012, Republic of Moldova, Europe
Managing Directors: Ieva Konstantinova, Victoria Ursu
info@omniscriptum.com

Printed at: see last page
ISBN: 978-620-8-64048-4

CONTENTS

1. introduction

Staphylococcus aureus are facultative anaerobic Gram-positive cocci, immobile, catalase and coagulase positive. The staphylococcal cell wall is resistant to isozyme and sensitive to isostaphin, which specifically cleaves the pentaglycine bridges common in this genus (LE LOIR et al., 2003). They are important hospital pathogens that cause a variety of clinical manifestations, including septicemia, septic wounds, septic arthritis, osteomyelitis, post-surgical septic shock syndrome, among others (BOYCE et al., 1997; VICENT et al., 1995). Most of the variation in this spectrum of clinical manifestations depends on the numerous virulence factors that each strain produces (LE LOIR et al., 2003). It is one of the main microorganisms causing hospital-acquired infections (MELZER, 2003), contributing significantly to the mortality and morbidity rate of patients, as well as being responsible for community-acquired diseases (COKILL. et al., 2004). Around 30 to 50% of the human population are carriers of this bacterium, and its main habitat is the mucous membrane of the nasopharynx. Here, these strains can persist as a transient or persistent member of the normal microbiota, without causing any symptoms. From the nasal microbiota, these pathogens can easily contaminate the hands of carriers and, in this way, contaminate food (FUEYO et al., 2000). *Staphylococcus aureus* has a variety of virulence factors, including adhesion molecules, peptide glycans, enzymes and extracellular toxins (ARCHER et al., 1998). The importance of *Staphylococcus aureus* as a pathogen lies in the combination of virulence mediated by its toxins, its invasive nature and its antibiotic resistance profile (LE LOIR et al., 2000). Genetic variation between strains of *Staphylococcus aureus* is being associated with their pathogenic potential (BOHACH et al., 1990; DAY et al., 2001). This variation occurs at the level of constitutive genes and accessory genes (LAN et al., 2000).

The ability of Staphylococcus aureus to evade the host's immune system is due to the existence of various virulence factors, including hemolysins, lipases, various proteases, exotoxins and enterotoxins (SOMERVILLE et al., 2001). Several *Staphylococcus aureus* accessory genes encode virulence factors, such as Enterotoxins (*sea - see; seg - ser; seu*) (FUEYO, et al.2005), Toxic Shock Syndrome Toxin (*tst*), Exfoliative toxins A and B (*eta* and *etb*), *leukocidins* (*pvl*, *lukE-lukD*) and surface-associated proteins such as Collagen Binding Protein (*cna*) (PROJAN, 1997). Some of these genes are often carried by mobile genetic elements, such as phages and Pathogenicity Islands (SaPIs), which are potentially mobile DNA segments of varying sizes, encoding genes associated with virulence (KAPPER and HACKER, 1999), being transferred horizontally between strains (*sea, tst, eta*) (BETLEY, et al.,1985; LINDSAY et al.,1998; YAMAGUCHI et al.,2000; YOSHIZAWA. et al.,2000).

In the hospital environment, MSSA strains are genetically diverse, but within this diversity, some strains are clearly related and probably represent successful lineages (MOORE et al., 2001).

In culture, *Staphylococcus aureus* only produces exotoxins at high bacterial density, in its post-exponential growth phase. In the pre-exponential phase, when at low bacterial density, the bacterium expresses surface molecules, such as fibronectin-binding proteins and fibrinogen-binding proteins, which allow the bacterium to adhere to and colonize host cells (Balaban et al., 2000).

According to BOHACH et al. (2000), *Staphylococcus aureus* exotoxins fall into two general groups: **membrane-active agents**, which include alpha, beta, delta, gamma and leukocidin toxins; and **toxins with superantigen activity (SAgs)**. Toxins with superantigen activity (SAgs) include the family of pyrogenic toxins (PTs), which are the staphylococcal enterotoxins (SEs), the toxic shock syndrome toxin (TSST-1) and the family of exfoliative toxins (ETs).

1.1 Toxins with membrane activity:

Staphylococcus aureus strains can produce four classes of cytotoxins, which comprise the four hemolysins, α, β, γ, and δ, as well as the Panton-Valentine leukocidin, designated as PVL.

1.1.1 Hemolysins

Hemolysins are cytotoxic agents whose action can aid in the invasive power of bacteria. The gene coding for alpha-toxin, *hla*, has been cloned and sequenced (SAKO, et al., 1983). **Alpha-hemolysin** is toxic to various mammalian cells and is also hemolytic to sheep erythrocytes, dermonecrotic and neurotoxic. These toxins cause pores to form in cell membranes, thus explaining their ability to cause lysis in erythrocytes. The alpha-toxin monomers attack the membranes and then form hexamers or heptamers, causing the pore to form (BHAKDI et al., 1991). It is reported that this pore formation induces the release of nitric oxide by endothelial cells and that alpha toxin induces apoptosis in lymphocytes (JONAS et al., 1994). Several studies have shown that alpha-toxin has a promoting effect on the onset of disease. In some studies, strains that did not produce alpha-toxin were less virulent than their isogenic ancestors that produced this toxin (BRAMLEY, 1989; CALLEGAN, 1994). Alpha-hemolysin is under the control of the *agr* regulatory gene and is produced in the post-exponential phase of bacterial growth. This toxin produces various effects on the host, most of which occur as a result of cellular osmotic imbalance due to the loss of ions through the pores formed (BHAKDI et al., 1991). An example of this is the activation of arachidonic acid metabolism in endothelial cells due to the influx of calcium. This leads to the formation of thromboxane and prostacyclins, which are vessel constrictors. Due to cell disruption secondary to osmotic imbalance, increased vascular permeability and pulmonary edema are observed (SUTTORP et al., 1985). As a result of the influx of calcium into the platelets caused by alpha-toxin, coagulation-activating factors are released (BHAKDI et al., 1988).

The gene for **beta-hemolysin** was cloned and sequenced by PROJAN et al. (1989). It is located on a 4Kb *Cla* I chromosomal DNA fragment. It is synthesized in the post-exponential phase of bacterial growth. Betahemolysin, unlike other cytolytic toxins, is an enzyme with sphingomyelinase activity (WADSTRON et al., 1972). The differences in the susceptibility of erythrocytes to beta-hemolysin can be explained by the amount of sphingomyelin present in the different erythrocytes. It has been observed *in vitro* that the action of beta-hemolysin is antagonistic to that of alpha-hemolysin. This is due to the action of beta-hemolysin on the erythrocyte membrane, thus preventing the polymerization of alpha-hemolysin. This suggests, therefore, that while red cells or any other cells containing sphingomyelin in their membrane represent a target for beta-hemolysin, alpha-hemolysin is then directed towards other targets.

Delta-hemolysin is a peptide containing 26 amino acids, which enhances the action of beta-hemolysin *in vitro*, increasing its hemolytic properties (RUZIKOVA, 1994). However, its role in virulence *in vivo* is still being studied. It is suggested that delta hemolysin has surfactant or channel-forming properties (BERNHEIMER, 1986).

Figure 1- Representation of the action of *S. aureus* alpha-hemolysin on the membrane of a eukaryotic cell:

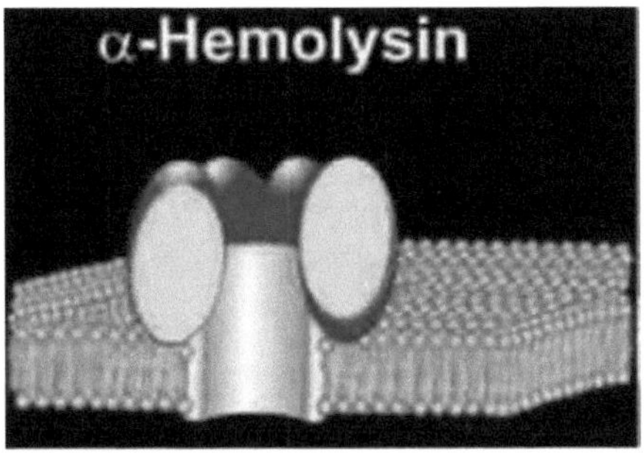

Source: FINK et al.; 1989

1.1.2 Bi-component toxins

Staphylococcus aureus strains produce two classes of **bicomponent toxins**, composed of two proteins with synergistic actions, secreted independently, referred to as class F and class S components. Four loci encoding bicomponent leukotoxins in *S. aureus* were characterized: *pvl*, *hlg*, *lukE-lukD* and *lukM-lukF'-PV*. Except for *hlgA*, all the other genes are co-transcribed in *tandem*, with the genes coding for proteins similar to the LukS-PV component (S component) being located *upstream of* the gene coding for the protein similar to LukF-PV (F component). Therefore, conventionally, the S class protein always corresponds to the first element mentioned.

The names S and F result from the speed (*Slow and Fast*) with which these proteins are eluted from a caboxy-methyl-cellulose ion exchange chromatography column (PREVÓST et al., 2002). These toxins comprise six S-class proteins, which are the first units to bind to the membranes of the target cells, and five F-class proteins, which correspond to the second element to continue the oligomer in formation. The interaction of these components induces a rapid activation and consequent opening of the calcium channels present in the membranes of the target cells, causing an influx of this cation into the cell. The oligomerization of the S and F components leads to the formation of heterohexamer pores on the surface of these cells, which are formed by an equimolar proportion of S and F proteins and are permeable to monovalent cations such as Na^+ , K^+ and etidium (WERNER et al., 2002). The bi-component toxins reported in this work comprise **Panton-Valentine leukocidin** (LukS-PV + LukF-PV, or simply **PVL**), **gamma-hemolysins** (HlgA + HlgB and HlgC + HlgB, where HlgA and HlgC are S components and HlgB is an F component) and LukE-LukD leukocidin. These toxins are synergistically toxic (in relation to the S and F components) to polymorph nuclear leukocytes, monocytes and macrophages in both rabbits and humans, and HlgA + HlgB can also lyse some T lymphocytes as well as human erythrocytes. The sensitivity of red cells varies according to the S/F pair formed. The leukocidin Luk E-D, on the other hand, is not strongly hemolytic on human erythrocytes, nor is it as leukotoxic when compared to other leukocidins (GRAVET et al., 1998). Any combination of an S-class protein with another F-class protein, heterologous or not, can be toxic (PREVÓST et al, 1995). An example of this is that the HlgA + LukD interaction strongly induces calcium channels (GRAVET et al.; 1998). These observations suggest both a difference in receptors and different forms of oligomerization of these toxins.

Studies indicate that almost all strains of *Staphylococcus aureus* produce gamma-hemolysins (approximately 99%) and only 2-3% of strains produce PVL. Studies conducted by PREVOST et al. (1995) indicate that *Staphylococcus aureus* strains have two separate *loci* for giving rise to gamma-hemolysins and PVL. Among the strains capable of producing both toxins, three S components (HlgA, HlgC and LukS-PV) and two F components (HlgB and LukF-PV) can be produced. These products show a high degree of similarity between their sequences. Three *open reading frames* (*ORF's*) comprise the genes for gamma-hemolysins, *hlg*, located on a 4.5 - Kb *locus* of a Sca-I chromosomal DNA fragment. The three *ORFs*, in order, are designated as *hlgA, hlgC and hlgB,* corresponding to the genes identified by COONEY et al (1993). The ORFs *hlgB* and *hlgC* are co-transcribed while *hlgA* is transcribed separately. The HlgB and HlgC toxins are typically referred to as hemolysins. The *locus* comprising *hlgA*, *hlgC* and *hlgB* is closely *related* to the *lukR locus*, encoding leukocidin R. PLOMMET M, et al.1(988) reported that *S. aureus 5R* strains could be an excellent source of gamma-hemolysins, since they do not produce alpha, beta or delta-hemolysin. PREVOST et al. (1995) also cloned genes encoding PVL, called *lukS-PV* and *lukF-PV*. The entire locus, *luk-PV*, was contained in an *EcoRV* chromosomal DNA fragment and consisted of two *ORFs*. The bacterial extracts of the clone exhibited leukotoxic activity, but no hemolytic activity. lukS-PV and *lukF* -PV were co-transcribed and were similar to *hlgC* and *hlgB*, respectively. No similarity to *hlgA* was found in the *luk-PV locus*. The acquisition of the *luk-PV locus is* attributed to the lysogenic action of bacteriophages (KANECO et al, 1997). However, horizontal transmission of this *locus* must be a very rare event, since neither the loss of this phenotype in toxin-producing strains nor natural dispersal between *different* isolates has been observed (PREVOST et al., 1995). With the three S components and two F components, it is possible for a given strain of *S. aureus* to produce six combinations of toxins, with one S component and one F component.

1.2 Toxins with superantigen activity

Several of the extracellular toxins produced by *Staphylococcus aureus* have superantigen properties, including enterotoxins, exfoliative toxins and Toxic Shock Syndrome Toxin (TSST-1). Superantigens affect the immune system by binding to major histocompatibility complex class II (MHC II) proteins and activate specific types of T-cells through the variable regions of the beta chains of T-cell receptors. As a consequence of the activation of these T-cells, cytokines are released, such as Tumor Necrosis Factor (TNF), various interleukins and interferon-gamma.

1.2.1 Staphylococcal Enterotoxins (SEs)

Staphylococcal enterotoxins (SEs) belong to the class of pyrogenic toxins with superantigen activity (PTSAgs), whose members share phylogenetic, structural and functional relationships with each other (BALBAN et al., 2000). Staphylococcal enterotoxin (*ses*) genes show a high percentage of identity in their nucleotide sequences (BETTLEY et al., 1992). Prior to the discovery of their superantigenic properties, enterotoxins were first recognized as potent emetic agents, and are now also involved in various allergic and autoimmune conditions such as *diabetes mellitus*, multiple sclerosis and rheumatoid arthritis (MCCORMICK et al.; 2001).

SEs function as potent gastrointestinal toxins and as superantigens, stimulating the proliferation of non-specific T cells (BALABAN et al., 2000). The emetic and superantigenic properties of these toxins are apparently distinct and are located on different protein domains (DINGES et al., 2000), indicating that they behave as bifunctional toxins (HARRIS et al., 1995). Although these are two distinct functions, located on different protein domains, there is a high correlation between these activities. In most cases, a decrease in superantigenicity activity (due to a genetic mutation) results in a decrease in enterotoxic activity (HARRIS et al., 1993). However, how these activities are linked remains unclear. One hypothesis is that enterotoxin activity may facilitate transcytosis, thus enabling the toxin to enter the bloodstream, thus allowing interaction with T cells, leading to superantigen activity (BALABAN et al., 2000). Of the pyrogenic toxins, SEs are the only ones with emetic activity (LE LOIR, et al., 2000).

Little is known about how SEs cause the symptoms of food poisoning. These toxins can have a direct effect on the intestinal epithelium and the vagus nerve, causing stimulation of the emetic center and increasing intestinal motility. A common finding in the structure of SEs is a cystine loop, which has been shown to be of considerable importance for emetic activity, based on analysis of mutants (DINGES et al.; 2000). SEI, on the other hand, does not have this cystine loop, so its emetic effect is considerably lower than that of other SEs (MUNSON et al., 1998).

The superantigenicity activity of enterotoxins results from the presence of enterotoxin receptors on the surface of antigen-presenting cells (APC). The term "superantigen" was introduced in 1989 by MARRACK & KAPPLER to describe a marked expansion of T cells, showing specific variable regions of the Beta chain (V-Beta) of the T cell antigen receptor (TCR), when exposed to these toxins

(MCCORMICK et al., 2001). These receptors are MHC class II molecules. Staphylococcal enterotoxin molecules therefore cross-link these MHC-II molecules with the variable region of the Beta chain (VB) of the T-lymphocyte antigen receptor (TCR). It is known that in the cellular immune response, the first event is the recognition of the antigen by the T-cell antigen receptor (TCR), anchored to the T-lymphocyte membrane. The antigen is presented to the TCR in the form of peptides linked to molecules of the major histocompatibility complex (MHC), class I and II, whose constituent molecules are membrane-bound proteins on the surface of antigen-presenting cells (APC). This type of recognition is the key to the high specificity of the immune response, where only a few T cells can recognize a specific antigen. Staphylococcal enterotoxins, because they behave like superantigens, interact with various T cells in an unspecific way, being a potent activator of these cells. To activate this action, these enterotoxins only require recognition of the V portion of the TCR Beta chain, thus promoting a link between the TCR and the class II MHC of the APCs. Conventional antigens, when presented together with the MHC on APCs, require recognition by all five TCR variable elements (V-Beta, D-beta, J-Beta, V-alpha and J-alpha). In contrast, SAGs are mainly recognized by the V-Beta portion of the TCR. In this way, they are able to stimulate around 20% of almost all T lymphocytes, while conventional antigens stimulate around 1:10000 T lymphocytes (MCCORMICK et al., 2001). This results in the synthesis of various cytokines, lymphokines, IFN and increased endotoxic shock, resulting in immunosuppression of T and B cells and impaired immune response to the invading agent (JARRAUD et al., 2001). The widespread expansion of T lymphocytes results in a massive release of cytokines by the host. These cytokines are believed to be responsible for most of the serious consequences of SAGs, which act as capillary vasodilators leading to hypotension, shock, multiple organ failure and death (MCCORMICK et al., 2001).

It has been proposed that enterotoxins trigger their emetic properties by binding to specific receptors in the intestine. This is confirmed by the fact that staphylococcal food poisoning (SFP), mediated by these toxins, whose main findings are vomiting, diarrhea and characteristic histopathological abnormalities, is not an infection, but an intoxication, in which the toxin had already been produced in the contaminated food, prior to ingestion (HOLMBERG, 1984). This allows us to speculate that the bacterium has selected a family of toxins that allows it to escape the hostile environment found in the digestive tract. Important properties shared by all SEs are their ability to induce vomiting and gastroenteritis after oral administration in primates, superantigenicity, similar conformations, pepsin stability and heat stability. The latter property means that the biological activity of the toxin remains unchanged even after heat processing of the food (HOLECKOVA et al., 2002).

The heterogeneity of SEs provides the criteria for their differentiation within the currently recognized classes (SEA - SEE; SEG - SER; SEU) (FUEYO, et al.; 2005), and SEC can be further differentiated into subclasses (SEC1, SEC2 and SEC3) (MARR et al. 1993). A staphylococcal exotoxin involved in TSS was initially named SEF (BERGDOLL et al.; 1981), but it does not show the biological activity, *in vivo,* of a true SE (FUEYO et al.; 2005). With regard to the similarity between the classes of SEs, 15% of the amino acid residues are completely conserved. Most of these conserved residues are located in the central and C-terminal portions of the sequences.

The genes coding for these enterotoxins have different genetic supports, most of which are mobile genetic elements. Not all SE genes are controlled by the *agr* system. Thus, *seb*, *sec* and *sed* are *agr-dependent*, while *sea and sej are* agr-independent (LE LOIR et al., 2000).

1.2.1.1 Staphylococcal Enterotoxin A (SEA):

SEA is the most common staphylococcal enterotoxin recovered from food poisoning outbreaks in the USA, accounting for 78% of all outbreaks (BALABAN et al., 2000). The ability of enterotoxin A to directly modify neutrophil function suggests an important role for this enterotoxin in the pathogenesis of some diseases (MULLARY, 2001). The gene for SEA (*sea*) is carried by a temperate bacteriophage. This gene is made up of 771 base pairs and encodes an enterotoxin A precursor of 257 amino acid residues. A portion of 24 N-terminal residues is processed, resulting in an active molecule. SEA is expressed in the exponential phase of growth, but is not regulated by *agr* (Tremaine et al.; 1993), unlike seb, sec and sed, which require a functional agr for maximum expression.

1.2.1.2 Staphylococcal enterotoxin B (SEB):

The coding region of the *seb* gene contains approximately 900 nucleotides. The precursor protein contains 267 amino acids. The *seb* gene *is* chromosomal in clinical isolates related to cases of food poisoning. However, in other strains it is located on a 750 KB plasmid.

1.2.1.3 Staphylococcal enterotoxin C (SEC):

SECs are a group of highly conserved proteins with important immunological cross-reactivities. There are three antigenically distinct SEC subtypes: SEC 1, SEC 2 and SEC3. The *sec3* gene contains 801 base pairs and encodes a precursor protein of 267 amino acids. The *sec3* gene *is* closely related to the *sec1* gene, with 98% identity between the nucleotide sequences. SEC3 differs from enterotoxins C1 and C2 by 4 and 9 amino acids, respectively. The *sec2* gene contains an *ORF* of 801 *bp*, encoding a precursor of 267 amino acids. The *sec* gene is located on a pathogenicity island (FITZGERALD et al., 2001).

1.2.1.4 Staphylococcal enterotoxin D (SED):

The *sed* gene is located on a 27.6 Kb plasmid called piB485. The product of the *sed* gene contains 258 amino acids. The mature polypeptide contains 228 amino acids. The superantigenicity activity of SED is dependent on Zn^{+2} for a high affinity with MHC class II.

1.2.1.5 Staphylococcal enterotoxin E (SEE):

The *see* gene encodes a 29-KDa protein that is apparently processed to a mature extracellular form with a molecular mass of 26 KDa. DNA sequence identity indicates that SEE, SED and SEA are strictly related. The *see* gene is carried by a defective phage (LE LOIR, 2000).

These toxins, known for their involvement in gastroenteritis, referred to as staphylococcal food poisoning (SFP), have received greater attention in recent years for being involved in Toxic Shock Syndrome (TSS) (BOHACH et al., 1990). Like other PTSAgs, SEs exhibit immune-modulatory and lethal properties necessary for TSS. SEs and TSST-1 share common structural and biological properties, suggesting that they derive from a common ancestor (MARRACK et al., 1990). The difference between SFP and TSS are well demonstrated in several studies and are a reflection of alternate routes of exposure to SEs.

1.2.2 Toxic Shock Syndrome Toxin (TSST-1)

Toxic Shock Syndrome (TSS) is a staphylococcal disease characterized by acute onset of high fever, diffuse erythematous *rash*, desquamation of the epidermis, hypotension, hypoalbuminemia, and involvement of three or more organ systems (REINGOLD et al., 1982). This syndrome was first described by TODD et al. (1978), when they reported this symptomatology in a group of seven children aged 8 to 17 with acute febrile illness. In the years that followed, there was an increase in the number of cases of TSS, and the syndrome was associated with young women during menstruation. Subsequently, it became evident that such cases also occurred in women who were not menstruating, and it became clear that this syndrome could affect any population (HERZER, 2001). It is now known that without adequate therapy, lethal shock can develop within 24 hours of the onset of symptoms (DEURENBERG, et al.; 2005). It is known that there are 3 **criteria for developing TSS**: (1) the patient must be infected or colonized by a strain of *Staphylococcus aureus*, (2) this strain must produce the TSST-1 toxin or similar toxins, (3) the toxin must find a route of entry into the circulation (HERZER, 2001). It was proposed that, in addition to the characteristic clinical findings, the **diagnosis of** TSS should incorporate **important laboratory findings**, such as: Isolation of S. aureus from a normally sterile mucosa or site; production of a toxin associated with TSS by the isolated strain; lack of antibodies related to the toxin in question during the acute phase of the disease; and development of antibodies against this toxin during the convalescence phase. This item, however, is not very appropriate since many patients do not develop neutralizing antibodies to this toxin (MCCORMICK et al.; 2001).

SST is one of the most common causes of death in intensive care units. TSS is separated into two distinct events: Menstrual and non-menstrual. In 1980, a clinical picture of epidemic TSS was described in women who used absorbent tampons, especially those with high absorbency. This is known to be due to the toxin's ability to cross mucosal barriers. The disease has been linked to toxigenic strains *of S. aureus*, mainly those of phage group I, located in the vaginal mucosa or uterine cervix. It is known that more than 99% of menstrual TSS cases are related to the use of tampons. Over 45% of all cases of TSS are non-menstrual (HERZER, 2001). Although non-menstrual TSS can occur in association with any staphylococcal infection, some main forms of this disease have been described. These include post-surgical TSS, TSS associated with influenza, erythematous syndrome and TSS associated with the use of diaphragms as a contraceptive method

(MCCORMICK et al.; 2001).

The **clinical presentation of** shock includes inflammation, acute onset of high fever, diffuse erythematous rash, vomiting, diarrhea, myalgias, circulatory failure and involvement of three or more organ systems (multiple organ dysfunction syndrome - MODS) (PARRILO et al., 1993). LPS, an endotoxin present in the outer membrane of Gram-negative bacteria, induces several of the findings associated with septic shock. LPS generally triggers septic shock through the over-activation of macrophages, which results in the over-production of cytokines, including TNF and IL-1 and IL-6. However, Gram-positive bacteria, especially Staphylococcus aureus, which lack an outer membrane and, moreover, do not produce LPS, are also capable of inducing septic shock. It is suggested that the superantigens produced by *Staphylococcus aureus* play an important role in septic shock, by stimulating the proliferation of T-cells and inducing the release of cytokines. The syndrome was first described by TODD et al. (1978) as the main systemic disease associated with non-invasive strains of *S. aureus*, and was designated TSS by these researchers. Patients with TSS usually do not have detectable bacteremia, although clinical findings of the disease are found, suggesting that TSS results from intoxication with the bacterial toxin (MCCORMICK et al., 2001).

TSST-1 was the first toxin related to TSS and is now accepted as being responsible for around 75% of clinical cases. It is encoded by a *tstH* gene (where H refers to human isolates), present on the chromosome. This gene can be found on one of the two pathogenicity islands, inserting itself either next to *tyrB* (SAPI1) or *trp* (SAPI2) (LYNDSAY et al., 1998). The *tst* gene was also found in the *SaPIbov* pathogenicity island. The main isolates from menstrual TSS contain *tstH* inserted close to the *trp* locus and are therefore considered to contain SaPI2 (MACCORMICK, et al.; 2001). It is responsible for approximately all cases of menstrual TSS, and non-menstrual TSS is also attributed to staphylococcal enterotoxins types B (SEB) and C (SEC) (MCCORMICK et al.; 2001). The toxin was characterized and identified as a new *S aureus* exotoxin by two researchers (SCHLIEVERT & BERGDOLL). TSST-1 is produced in around 100% of *S. aureus* strains isolated from the cervical or vaginal mucosa of women with SST, and in approximately 50% of isolates from other body sites of patients with non-menstrual SST. Only TSST-1 was involved in cases of menstrual TSS, due to TSST-1's ability to cross mucosal barriers, unlike enterotoxins (SCHLIVIERT et al.; 2000). TSST-1 induces a humoral response in vivo which is believed to be protective. Around 90 to 95% of middle-aged women have detectable antibody titres against this toxin. Patients with TSS produce an ineffective response to TSST1, with titers below 1:5, while healthy patients have titers above 1:100 (HERZER, 2001).

1.2.3 Exfoliative toxins:

Two serologically distinct exfoliative toxins have been described, called exfoliative **toxin A** and **exfoliative toxin B** (**ETA** and **ETB**, respectively). Both genes are designated as *etA* and *etB*, and their sequences have already been determined (LEE et al.,1987). Among the clinical manifestations caused by ET-producing strains is staphylococcal scalded skin syndrome (SSSS), which can manifest in its classic form, generalized throughout the epidermis, or localized, characterizing impetigo. Impetigo can be bullous or non-bullous (YAMASAKI et al., 2005). The scalded skin syndrome (SSSS), usually observed in newborns (Ritter's disease), generally in the form of hospital outbreaks (HELLALI et al., 2004), is an exfoliative dermatitis mediated by infection with strains *of Staphylococcus aureus* that produce both ETA and ETB. However, the presence of the *etb* gene is strongly associated with generalized SSSS, while the *eta* gene is associated with bullous impetigo (YAMASAKI et al., 2005). Currently, the action of these toxins has been linked to serine proteases (Dancer et al., 1990). These toxins trigger an intradermal separation of the stratum granulosum by specifically cleaving desmoglein-1, which is a desmosomal protein that mediates intercellular adhesion of the keratinocytes present in the granular layer (YAMASAKI et al., 2005). In SSSS, when blisters are observed, the intradermal fluid is sterile, and the infecting strain is usually recovered from sites distant from the lesion, such as the throat or nasal cavity (YAMASAKI et al.; 2005). Some authors attribute a strong predominance of ETB in this syndrome, in relation to ETA (KANZAKI et al.; 1997). This condition is typical of children, although it can also occur in adults (LISA, 2004). This disease starts abruptly with fever and erythema, usually near the mouth, and can spread to the whole body within a few days, followed by the formation of sterile blisters on the epidermis, which cause the epidermal layers to separate, especially at the level of the stratum granulosum (AMAGAI et al., 2002). Exfoliation of the epidermis, which predisposes the patient to secondary infections by opportunistic pathogens (LADHANI et al.; 1999), occurs at a site far from the infection site. It is believed that SSSS originates through the absorption of ETs produced by strains from these distant sites. These toxins reach the systemic circulation by diffusion, in order to reach the target sites (YAMASAKI et al.; 2005). According to LADHANI (2003), the high incidence of generalized SSSS in children is due to the less effective renal *clearance of* these toxins and the immune immaturity observed in this population, resulting in low titers of anti-ET antibodies. Mortality in children is around 3%, when appropriate antibiotic therapy is used, and is usually associated with secondary infections (GEMMELL, 1995). However, hospital outbreaks continue to be reported (MACKENZIE, et al., 1995). In contrast, in bullous impetigo, which is a localized infection also caused by ET-producing strains of S. aureus, the liquid is formed by purulent exudate, and these strains are usually isolated from this liquid. This clinical manifestation can occur at any age (YAMASAKI, et al; 2005). Koning et al. demonstrated the relationship of ETB-producing strains with PVL and a plasmid resistant to various drugs, called pSK41 (Gillet et al.; 2002). These authors reported that the relationship between these virulence factors contributes to the severity of non-bullous impetigo.

ETA is encoded by a chromosomal gene (*eta*) within a 43 Kb segment of a temperate phage (YAMAGUCHI et al., 2000) and is the most common of the exfoliative toxins. ETB, on the other hand, is encoded by a plasmid of approximately 38.2 to 38.5 Kb. ETA and ETB share a significant identity between their amino acid sequences and also have important biophysical properties in common (LEE et al., 1987). ETA is a very stable toxin, which is resistant to extreme heat, while ETB is sensitive to heat (LISA, 2004). Recently, a new exfoliative toxin, ETD, was identified from a strain isolated from wounds. It was initially related to an *ORF* with significant homology to known ETs (YAMAGUCHI et al., 2002.). A fourth exfoliative toxin, potentially active in humans, called ETC, was isolated from infections in horses, and showed activity in neonatal rat models (SATO et al., 1994), but has not yet been associated with disease in humans. ETs, like most staphylococcal proteins associated with virulence, are encoded in mobile genetic elements (NOVICK, 2003), allowing the horizontal transfer of these virulence factors. For a reason that is not yet well understood, exfoliative toxins are not co-produced with TSST-1 by *Staphylococcus aureus* strains, although Ets can be co-produced with other enterotoxins (GRAVET et al., 2001). Like most *Staphylococcus aureus* virulence factors, TEs are regulated by the *agr* locus (NOVICK, 2003).

There are various controversies surrounding the association of ETs as superantigens. Although erythema, fever and purulent foci can be found in SSSS, they are not always present. Although specific T-cell stimulation has been demonstrated, SSSS generally lacks the systemic symptoms associated with the superantigen-mediated syndrome. However, superantigenicity is based on the fact that the bacterium is not necessarily found in all bullous lesions present in SSSS (VATH et al., 1999).

1.3 *Locus agr*

The expression of exotoxin genes is controlled by regulators such as *agr* (accessory gene regulator) (PROJAN, 1997), with four different and mutually inhibitory classes of this gene being identified (JARRAUD et al.,1997; JI et al.,1997.). The *agr* locus is a positive regulator for exotoxins secreted during the post-exponential phase of growth, but a negative regulator for surface-associated virulence factors. Exotoxins regulated by *agr* are not synthesized or are significantly reduced in agr-defective mutants (RECSEI et al., 1986). The production of alpha-hemolysin, delta-hemolysin and TSST-1 appears to be strongly regulated by agr. These toxins are produced in high quantities by *agr* strains+ and are produced in minimal quantities in *agr-* strains, and are referred to as **class-I toxins**. On the other hand, beta-hemolins, enterotoxins B and C and exfoliative toxin A do not seem to be as strongly regulated by *agr*, since these proteins are produced in high quantities in *agr+* strains and in moderate quantities in *agr-* strains, and are referred to as **class II toxins**. **Class III proteins** are protein A, coagulase and fibronectin-binding protein, which are produced in high concentrations in *agr-* strains, but in low concentrations in *agr+* strains. The *agr locus* has been sequenced and cloned, comprising at least five genes: *agrA*, *agrB*, *agrC*, *agrD* and the gene for delta-hemolysin (*hld*) (PENG et al., 1989). The *agr locus* appears to be a two-component regulatory system, with *agrD* originating a signaling component and *agrA* encoding a transcriptional activator. The two main transcripts produced by agr are designated as RNAII (agr A, B, C, D) and RNA III (*hld*). RNA III acts primarily to initiate the transcription of virulence factors.

The genes coding for virulence factors addressed in this work are the following: enterotoxins A, B, C, D and E (*sea, seb, sec, sed see*); Toxic Shock Syndrome Toxin (*tst*); Exfoliative Toxins A and B *(eta and etb*);

2. Material and methods

2.1 Bacterial samples:

This study used 37 isolates of *Staphylococcus aureus* from 12 laboratories in Brazilian public and private hospitals. These isolates were confirmed as oxacillin-sensitive *S. aureus* (**OSSA**) by *screening* for oxacillin. Of these 37 OSSA, seven were isolated from foods blamed for cases of food-borne infections. The choice of samples was based on the heterogeneity of clinical material and food from cases of food poisoning.

Table 1 - Distribution of the strains studied according to origin and source of isolation.

n	Strains	Origin	Material
01	C1	Hosp. A	LCR
02	C2	Hosp. A	LP
03	C3	Hosp. A	Sg
04	C4	Hosp. A	LCR
05	C5	Hosp. A	LP
06	C6	Hosp. B *Sg*	
07	C7	Hosp. B	Sg
08	C8	Hosp. C	Sg
09	C9	Hosp. C	Sg
10	C10	Hosp. C	Sg
11	C11	Hosp. D	Sg
12	C12	Hosp. D	Sg
13	C13	Hosp. and	Sg
14	C14	Hosp. and	Sg
15	C15	Hosp. F	Sg
16	C16	Hosp. F	Sg
17	C17	G	Sg
18	C18	Hosp. H *Sg*	
19	C19	Hosp. H	Sg
n	**Strains**	**Origin**	**Material**
20	C20	Hosp. H	Sg
21	C21	Hosp. H	Sg
22	C22	Hosp. I	Sg
23	C23	*J*	Sg

24	C24	L	Sg
25	C25	L	Sg
26	C26	L	Sg
27	C27	L	Sg
28	C28	M	Sg
29	C29	N	Sg
30	C30	N	Sg
31	Al-1	L-1	Al-
32	Al-2	L -2	Al-
33	Al-3	L -3	Al-
34	Al-4	L -4	Al-
35	Al-5	L -5	Al-
36	Al-6	L -6	Al-
37	Al-7	L -7	Al-

***Hosp** A - N: Hospitals in different Brazilian cities where the strains were isolated

** L1 - 7: Food microbiology laboratory where 7 strains were isolated from food related to food toxi-infections (annex-1)

*** CSF - Cerebrospinal fluid; LP - Pleural fluid; Sg - Blood; Al - Foods related to food toxi-infections.

2.2 Screening to confirm sensitivity to oxacillin:

All the isolates were screened for oxacillin sensitivity using the method described by the NCCLS (2005), which uses Mueller-Hinton agar containing 6µg/mL of oxacillin and 4% sodium chloride (NaCl). The inoculum was applied to the plate using a swab soaked in a bacterial suspension equivalent to McFarland's 0.5 scale. The plate was incubated at 35°C for exactly 24 hours. The observation of one or more colonies in the area where the inoculum was applied defined the result as positive, i.e. the strain was considered resistant to oxacillin.

2.3 Analysis of DNA polymorphism by pulsed field electrophoresis (PFGE)

2.3.1 Extraction of genomic DNA for PGFE

The *Staphylococcus aureus* strains were seeded on 5% sheep's blood agar (AS) (OXOID, England) and incubated at 37°C for 24 hours. They were then plated onto soy-trypticasein broth (TSB) (DIFCO, USA) and incubated at 37°C for one night. After this period, the samples were centrifuged at 3,345g for 10 minutes. The sediment was resuspended in 1mL of sterile saline solution and transferred to a microcentrifuge tube which had previously been weighed. This suspension was centrifuged at 23,100□g for 15 minutes and the supernatant was discarded. Next, the sediment was resuspended in 1mL of sterile saline solution, centrifuged at 23,100g for 15 minutes and the supernatant discarded. This procedure was repeated three times. The sediment was then weighed and re-suspended in 50mM EDTA pH8.0, to a final concentration of 100µg of bacterial mass per µL.

From this suspension, 15µL was transferred to a micro-centrifuge tube containing 215µL of EC buffer (Appendix II), plus 250µL of 2% low melting point agarose (Sigma, USA) and 20µL of lysostaphin solution (100µg/mL) (Sigma, USA). This suspension was transferred to the molds where it remained for 30 minutes at 4°C for the blocks to solidify. After this stage, the blocks were transferred to a 24-hole cell culture plate containing 2mL of EC buffer in each hole and incubated at 37°C for five hours under gentle agitation. After incubation, the blocks were washed with 2.5mL of CHEF TE buffer (Appendix II), then incubated with 2mL of ES buffer (Appendix II), plus 100µL of proteinase K solution (20mg/mL) (GibcoBRL, Life Technologies, USA), for one night at 50°C. The blocks were then washed with 2.5mL of CHEF TE buffer (4 washes with 1-hour incubations each) and stored in this solution until they were subjected to enzymatic digestion and electrophoresis (PFALLER, 1993).

2.3.2 Enzymatic digestion

The chromosomal DNA was digested using the *SmaI* restriction enzyme (Amersham Pharmacia Biotech, USA), which recognizes and cleaves nucleotide sequences that appear only a few times along the DNA of *Staphylococcus aureus* (12 to 20 restriction sites) (PREVOST et al., 1992; STRUELENS et al., 1992). For the treatment with restriction enzymes, an agarose block from each sample was transferred to a 96-hole cell culture plate containing 300µL of DNS buffer. The DNS buffer was then removed and 300µL of DNS buffer was added again, and the blocks were incubated at room temperature for 1 hour. This process was repeated 4 times. After this period, the DNS buffer was replaced with 200µL of restriction enzyme buffer and the block was incubated at room temperature for 2 hours. The buffer was then discarded and 200µL of enzyme buffer and 40µL of *Sma* I enzyme were added and the blocks were then incubated for 2 hours at 4°C and then for 15 hours at 30°C (PFALLER, 1993).

2.3.3 Pulsed Field Electrophoresis:

Pulsed field electrophoresis was carried out on a 1% agarose gel (Sigma, USA) in 0.5%TBE (90mM Tris, 90mM boric acid, 2mM EDTA [Amersham Pharmacia Biotech, USA]) (SAMBROOK et al., 1989), in CHEF DRII System equipment (BIO-RAD, USA) with pulse time intervals of 1 to 30 seconds for 24 hours, angle of 120°, temperature of 14°C and voltage of 6.0 volts/cm (PFALLER, 1993).

After separation, the gels were stained with an aqueous solution containing 0.5µg/mL of ethidium bromide (Sigma, USA) and photographed under transillumination with ultraviolet light. The molecular weight standard used was Lambda DNA ladder PFG (New England Biolabs, Inc., USA), with band extensions from 48.5Kb to 1.0Mb (PFALLER, 1993). A strain of S. aureus ATCC 25923 was used as a control. The bands obtained by electrophoresis were analyzed visually according to the guide for interpreting pulsed field electrophoresis proposed by TERNOVER et al. (1995). (Table 2).

Table 2- Criteria for interpreting pulsed field electrophoresis (PFGE)

Categories	No. of genetic differences	No. of different fragments	Interpretation

Similar	0	0	Strains are part of the same clone
Related	1	2-3	strains are considered as a subgroup
Possibly related	2	4-6	possibly related strains
Different	Greater than or equal to 3	greater than or equal to 7	different strains

Source: TERNOVER et al.,1995

2.4 Obtaining genomic DNA for amplifications:

Total DNA was extracted using the phenol-chloroform extraction technique (SAMBROOK et al., 1989). The strains were grown *overnight* in LB broth and resuspended in lysis buffer (50nM Tris, pH 8, 5mM EDTA pH 8, 50mM NaCl). Lysostaphin (final concentration 20mg/L) was used for the cell rupture phase. Lysozyme and proteinase K were used to lyse the proteins, which were removed from the medium by phase separation with phenol-chloroform-isoamyl alcohol. The remaining DNA in the aqueous phase was precipitated with ice-cold absolute ethanol and centrifuged in a Sorvall RC2-B (SS34 rotor), adhering to the tube wall. It was then solubilized in TE buffer (10mM Tris, 1 mM EDTA, pH 8) and stored at -20°C.

The quantification of the DNA and its integrity were verified by electrophoresis in a 0.8% agarose gel, compared with a known molecular mass standard (*High DNA Mass™ Ladder - Invitrogem™*). The gel was stained with ethidium bromide solution (*Invitrogem™*) at a final concentration of 0.5 µg/mL and its bands were visualized under ultraviolet (UV) light.

2.5 Detection of genes coding for virulence factors and the type of *agr* present:

2.5.1 PCR for the detection of the genes of the main virulence factors:

Thirteen pairs of primers were used to amplify specific structures for each of the 13 genes related to the virulence factors investigated (table 4). Standard strains containing the genes encoding the virulence factors investigated were used as internal controls.

Table 3- Control-positive strains for each toxin gene investigated

Toxin gene	*eta*	*etb*	*hla,hlb,hlg*	*tst*	*sea*	*seb*	*sec*	*sed*	*pvl*	**LukE D**
Standard strain	ZM	N5	N315	N315	Mu50	T116	N315	RN4220	MR108	N315

eta- *Gene coding for Exfoliative Toxin A;* ***etb***- *Gene coding for Exfoliative Toxin B;* ***hla*** *- Gene coding for alpha-hemolysin;* ***hlb*** *- Gene coding for beta-hemolysin;* ***hlg*** *- Gene coding for gamma-hemolysin;* ***tst*** *- Gene coding for TSST-1;* ***sea*** *- Gene coding for enterotoxin A;* ***sec*** *- Gene coding for enterotoxin C;* ***sed*** *- Gene coding for enterotoxin D;* ***pvl*** *- Gene coding for Panton-Valentine Leukocidin;* ***LukED*** *- Gene coding for leukocidin E-D.*

The ZM, N5, RN4220 and MR108 strains were donated by Professor Teruyo Ho and Dr. Keiichi

Hiramatsu, from Juntendo University, Tokyo, as carriers of the genes mentioned in Table 3. The Mu50 strain was cited by KURODA et al., 2001, as carrying the *sea* gene. The N315 strain had its genome published in the *Lancet*, under reference: 357 (9264), 1225-1240 (2001). The T116 strain had its amplicon for the *seb* gene sequenced at the human genome study center at the Biosciences Institute of the University of Sao Paulo.

Table 4 - Initiators used for the toxins under study

Toxin	Gene	Initiator	Nucleotide sequence
SEA	*sea*	SEA-1	5' - GAAAAAAGTCTGAATTGCAGGGAACA - 3'
		SEA-2	5' - CAAATAAATCGTAATTAACCGAAGATTC - 3'
SEB	*seb*	SEB-1	5' - ATTCTATTAAGGACACTAAGTTAGGGA - 3'
		SEB-2	5' - ATCCCGTTTCATAAGGCGAGT - 3'
SEC	*sec*	SEC-1	5' - GTAAAGTTACAGGTGGCAAAACTTG - 3'
		SEC-2	5' - CATATCATACCAAAAAGTATTGCCGT - 3'
SED	*sed*	SED-1	5' - GAATTAAGTAGTACCGCGCTAAATAATATG - 3'
		SED-2	5' - GCTGTATTTTTCCTCCGAGAG - 3'
SEE	*see*	SEE-1	5' - CAAAGAAATGCTTTAAGCAATCTTAGGC - 3'
		SEE-2	5' - CACCTTACCGCCAAAGCTG - 3'
TSST-1	*tst*	TST-1	5' - TTCACTATTTGTAAAAGTGTCAGACCCACT - 3'
		TST-2	5' - TACTAATGAATTTTTTTATCGTAAGCCCTT - 3'
ETA	*eta*	mpETA-1	5' - ACTGTAGGAGCTAGTGCATTTGT - 3'
		mpETA-3	5' - TGGATACTTTTGTCTATCTTTTTCATCAAC - 3'
ETB	*etb*	mpETB-1	5' - CAGATAAAGAGCTTTATACACACATTAC - 3'
		mpETB-2	5' - AGTGAACTTATCTTTCTATTGAAAAACACTC - 3'
Toxin	**Gene**	**Initiator**	**Nucleotide sequence**
PVL comps S and F	*lukS-PV*	PVL-1	5' ATCATTAGGTAAAATGTCTGGACATGATCCA- 3'
	lukF-PV	NPVL-2	5' - GCATCAASTGTATTGGATAGCAAAAGC - 3'
LukE-LukD	*lukE-lukD*	LUKED-1	5' - TGAAAAAGGTTCAAAGTTGATACGAG - 3'
		LUKED-2	5' - TGTATTCGATAGCAAAAGCAGTGCA - 3'
HLA	*hla*	HLA-1	5' - CTGATTACTATCCAAGAAATTCGATTG - 3'
		HLA-2	5' - CTTTCCAGCCTACTTTTTTATCAGT - 3'
HLB	*hlb*	HLB-1	5' - GTGCACTTACTGACAATAGTGC - 3'

		HLB-2	5' - GTTGATGAGTAGCTACCTTCAGT - 3'
HLG	*hlg*	mpHLG-1	5' - GTCAYAGAGTCCATAATGCATTTAA - 3'
		mpHLG-2	5' - CACCAAATGTATAGCCTAAAGTG - 3'

Initiators synthesized according to JARRAUD et al., 2002

SEA - SEE: Enterotoxins A, B, C, D, E; **TSST-1**: Toxins associated with Toxic Shock Syndrome; **ETA**: Exfoliative Toxin A; **ETB**: Exfoliative Toxin B; **PVL**: Panton-Valentine Leukocidin; **LukE-LukD** : leukocidin E-D; **HLA**: alpha-hemolysin; **HLB**: Beta-hemolysin; **HLG**: Gamma-hemolysin.

Note: Accompanying each toxin is its coding gene, always represented in italics and lowercase.

To amplify the gene structures corresponding to the toxin genes under study, the following protocol was used as described by JARRAUD *et al* (2002), with a prior dilution of 1:100 of the DNa obtained. A volume of 20µL of a reaction mixture containing 2µL of DNA, 0.5 units of Taq polymerase and a final concentration of the following components was used: 200µM of each dNTP; 1.5mM of $MgCl_2$; 75mM Tris-HCl pH 9.0; 20mM $(NH_4)_2SO_4$; 0.01% Tween 20; 250nM of the *primers*. Thirty amplification cycles were run, each consisting of 60s at 95°C, 60s at 55°C and 2 minutes at 72°C. Each PCR test was controlled with positive controls of standard strains with the virulence genes under study and negative controls with microorganism-free samples. Ten microliters of PCR product was separated by electrophoresis on a 0.8% agarose gel and visualized under ultraviolet light after staining with ethidium bromide.

2.5.2 . Determination of the polymorphism profile of the *agr locus:*

Table 5 - Initiators used to classify the agr locus:

Initiator	Nucleotide sequence
PAN-agr	5' - ATGCACATGGTGCACATGC - 3'
agr-I	5' - GTCACAAGTACTATAAGCTGCGAT - 3'
agr-II	5' - TATTACTAATTGAAAAGTGGCCATAGC - 3'
agr-III	5' - GTAATGTAATAGCTTGTATAATACCCAG - 3'

Initiators synthesized according to JARRAUD et al., 2002

To amplify the *agr locus* under study, the folowed PAN- agr primer was used, which anneals to a region common to all the polymorphic agr types described. With this primer, each of the reverse primers agrI, agrII and agrIII were used in each reaction, which anneal to segments specific to each of the respective polymorphic types I, II or III, respectively. A volume of 20µL of a mixture containing 2µL of DNA (diluted 1:100), 0.5 units of Taq polymerase and a final concentration of the following components were used for the PCR reaction: 200µM of each dNTP; 1.5mM of $MgCl_2$; 75mM Tris-HCl pH 9.0; 20mM $(NH_4)_2SO_4$; 0.01% Tween 20; 250nM of the *primers*. Thirty amplification cycles were run, each consisting of 60s at 95°C, 60s at 55°C and 2 minutes at 72°C. Each PCR test was controlled with positive controls of standard strains with the subtypes of the *agr locus* under study and negative controls with microorganism-free samples. Ten microliters of PCR product was separated by electrophoresis on a 0.8% agarose gel and visualized under ultraviolet light after staining with ethidium bromide. Control strains carrying the main types of *agr* isolated in Brazil were used as a positive control.

Figure 2- Representation of the *agr A,* C, D and B genes and the regions used to amplify the polymorphic segments of agr types 1, 2, 3 and 4.

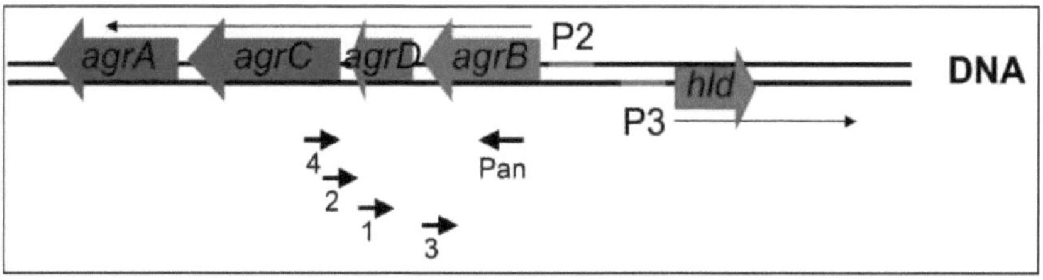

"from McCulloch et al. in XXIII Brazilian Congress of Microbiology, 13/1240-2, 2005"

Table 6 - Standard strains used to classify the polymorphism of the agr locus

agr

Type agr	I	II	III
strain	R137	N315	ATCC 25923

Agr type confirmed by sequencing in the genome project, IB, USP.

2.6 Agarose gel electrophoresis:

For visualization and analysis of the DNA fragment amplified in the PCR reaction, an agarose gel was prepared at 0.8% in a 0.5 X TBE cap (Trisma base - 890 mM, EDTA - 25 mM, boric acid - 89 mM). To check the sizes of the PCR products, aliquots of these samples, as well as a known molecular mass standard (1Kb *DNA Ladder* - Invitrogem™), mixed with a buffer solution (0.25% bromophenol blue; 0.25% FF xylene cyanol; 30% glycerol) were subjected to electrophoretic separation for about 45 minutes at 100 volts. The gel was stained and visualized as described in section 3.4.

2.7 Determination of the size of the amplified DNA fragments:

To obtain the image of the agarose gel (previously stained with ethidium bromide), both photographically and digitally, the following system was used

Photodocumentation System Model: DP-001.FDC Version 10 (*Vilber Loumart, Marne-lavallée Cedex 1* - France). The photographic image was printed using the *Video Graphic Printer* UP-895CE (SONY).

After obtaining the digital image, the sizes of the DNA fragments were analyzed using *PhotoCaptMw*

Version 10.01 for *Windows*.

Panton-Valentine Leukocidin (*pvl*); Leukocidin E-D (*luk-E-lukD*); and the hemolysins: Alpha, Beta and Gamma (*hla, hlb, hlg).* In addition to these genes, the type of *agr locus* present in each strain will be studied.

Because of its important role in controlling the virulence genes of *Staphylococcus aureus*, this gene has now become an important target for therapy, since its response can be controlled by synthetic peptides (MAYVILLE et al., 1999). As these responses are agr-specific and as the distribution of agr types in clinical strains is poorly understood, it is important to know the prevalence of the agr type in nosocomial strains.

3. Results

2.8 Screening for oxacillin sensitivity

All the isolates used in this study were characterized as OSSA when they were screened for oxacillin resistance using the method described in section 3.2.

2.9 Analysis by Pulsed Field Electrophoresis (PFGE):

The profile obtained from pulsed field electrophoresis of the OSSA strains in this study shows a diversity of clonality between isolates from the same hospital. Exceptions were isolates C3 and C5 from hospital A, which showed only one difference band, and isolates C31 and C33 from hospital N, which showed two different bands. These exceptions could therefore be related to the same clone, according to the criteria presented by TERNOVER et al (1995). The C10, C11 and C4 strains did not have their molecular profiles analyzed due to difficulties in recovering them.

Figure 2 - Photograph of the gel obtained from pulsed field electrophoresis (PGFE):

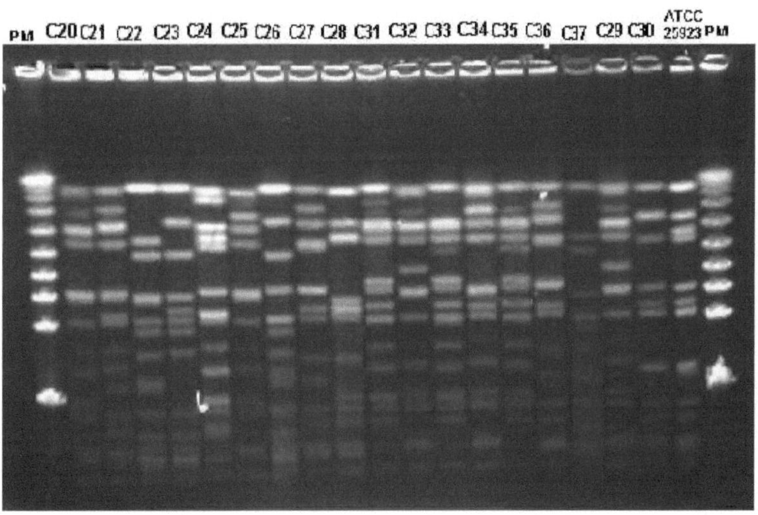

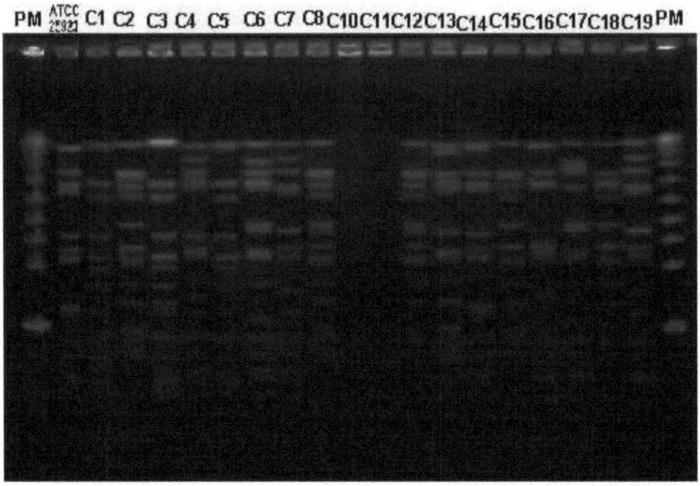

3.3 Determining the *agr* type

Figure 3: Fragments obtained from the amplification of agr types I, II and III

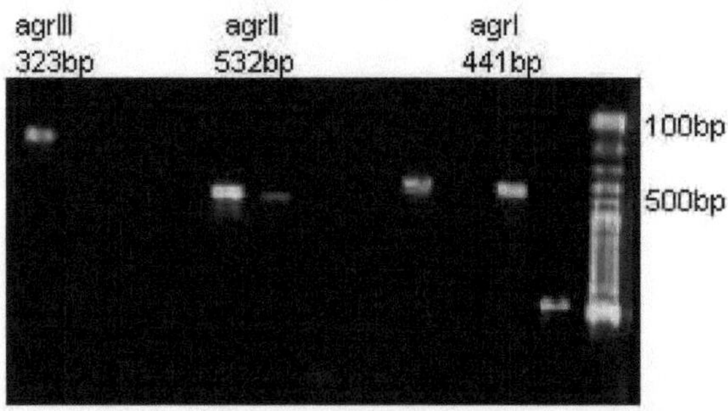

Table 7 - Distribution of the *agr* type found in the 37 OSSA isolates studied

agr	*OSSA strains*	*Total (%)*
I	**C2,C8,C10,C13,C25,C36,C29**	7 (19)
II	C9,C11,C14,C24,C28,C37,C30	7 (19)
III	C1,C3,C4,C5,C6,C7,C12,C15,C16,C17,C18,C19,C20, C21,C22,C23,C26,C27,C31,C32,C33, C34,C35	23 (62)

The acronyms and numbers that accompany each isolate are just codes to identify them.

The percentage figures are in brackets;

OSSA (Oxacillin Sensitive *Staphylococcus aureus*);

Table 8 - Source of isolates according to the type of *agr* found

Source of **Isolation** *n=37*	***agr***		
	I (%)	***II (%)***	***III (%)***
Al	1(14)	1(14)	5(72)
Clinical Mat	6(20)	6(20)	18(60)

n - corresponds to the number of strains related to each item;

the percentage values are in brackets;

Clinical Material: Strains from clinical material;

Al: strains from foods involved in food toxi-infections

3.4 Determination of the genes encoding virulence factors:

Figure - 4: Fragments obtained from the amplification of some of the virulence factors studied

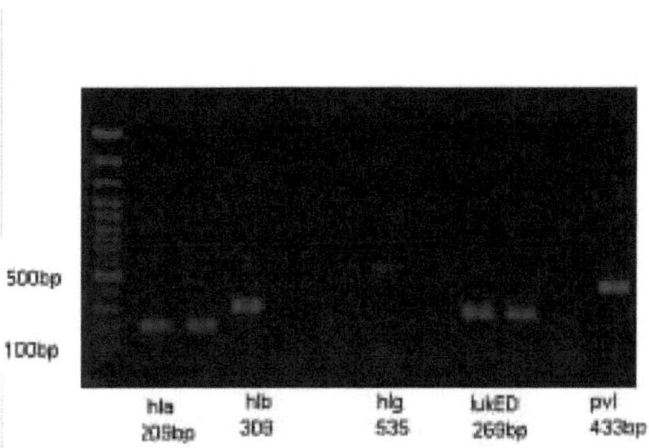

Figure 4-a: Amplified fragments of the *hla*, *hlb*, *hlg*, *lukED* and *pvl* genes

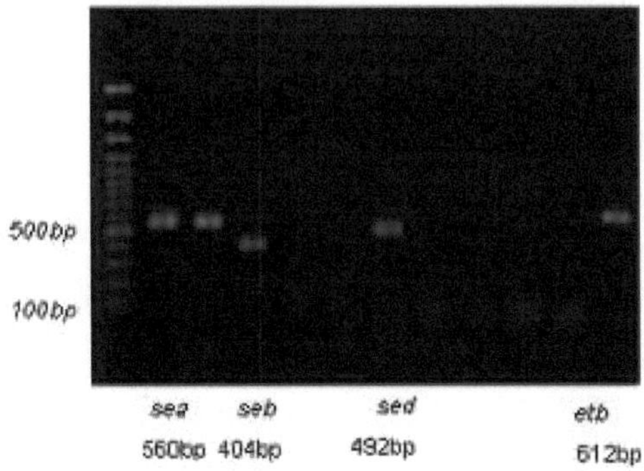

Figure - 4-b: Amplified fragments of the *sea*, *seb*, *sed*, and *etb* genes

3.4.1 Table 9 - Toxin coding genes according to the type of *agr* found

Toxins	***agr***		
(n=37)	**I (%)**	**II (%)**	**III (%)**
sea	0	0	**15** (100)
seb	**1**(14)	**1**(14)	**5**(72)
sec	0	**1** (9)	**11**(91)
sed	0	**1**(50)	**1**(50)
see	0	0	0
tst	**1** (11)	**2** (22)	**6** (67)
eta	0	0	0
etb	0	0	0
pvl	**1**(14)	**1**(14)	**5** (72)
luKED	**6** (17)	**7** (20)	**22** (63)
hla	**7** (18)	**7** (18)	**23** (64)
hlb	**5** (17)	**4** (13)	**20** (70)
hlg	**7** (19)	**7** (19)	**22** (62)

n corresponds to the number of isolates related to each item; percentage values are in brackets;

2.9.1 Table 10 - : Frequency of toxin-encoding genes

Genes encoding exotoxins	**Frequency (%)**
sea	*40*
seb	*19*
sec	*34*
sed	*5*

see	*0*
tst	*25*
eta	*0*
etb	*0*
pvl	*19*
luKED	*94*
hla	*100*
hlb	*78*
hlg	*97*

n corresponds to the number of positive strains for each gene;

2.9.2 Table 11 - : List of toxin genes found according to the origin of the strains studied

Exotoxin coding genes n=37	**Food n=7**	**Clinical Staff n=30**
sea	*4 (57)*	*11(37)*
seb	*0*	*7(23)*
sec	*3 (43)*	*9 (30)*
sed	*0*	*2 (6)*
see	*0*	*0*
tst	*2 (28)*	*5 (16)*
eta	*0*	*0*
etb	*0*	*0*
pvl	*1 (14)*	*6 (20)*
luKED	*7(100)*	*18 (60)*
hla	*7 (100)*	*30 (100)*
hlb	*6 (9)*	*23 (77)*
hlg	*7(100)*	*29 (97)*

n corresponds to the number of strains related to each item; Percentage values are in parentheses; **CSF** (Cerebrospinal Fluid); **PL Liq** (Pleural Fluid); **Hec** (Blood Culture)

4. Discussion

Microbial pathogens have a great versatility in colonizing different ecological niches and causing infections in different anatomical sites of their hosts. The molecular process responsible for the disease in specific hosts is little known. However, the genetic variation between strains is of great importance in this context (FITZGERALD, et al.; 2001).

Staphylococcus aureus can be a commensal as well as an extremely versatile pathogen in humans. The disease caused by this bacterium can be the result of direct tissue invasion or due to the action of a variety of exotoxins released. Pathogenic *Staphylococcus aureus* strains express several virulence factors, which include both secreted toxins (exotoxins) and cell surface molecules associated with bacterial adhesion. At least 34 different extracellular proteins are produced by these strains, several of which already have a defined role in the pathogenesis of staphylococcal disease (LISA, 2004). With the exception of toxemia, the molecular basis of the pathogenicity *of Staphylococcus aureus* is multifactorial, depending on the presence and expression of various accessory genes. In invasive processes, there are several virulence factors that *S. aureus* strains can synthesize, including capsular polysaccharides, surface-associated proteins and various exotoxins (SUTRA et al., 1993).

Genetic-molecular methods are capable of detecting the **potential of** strains to produce the toxin in question, as there are cases in which toxin genes are not expressed (Koleckova, et al., 2002). Therefore, the findings of this study provide us with data on the genetic aspects of the intrinsic virulence factors in the *Staphylococcus aureus* strains studied. We therefore analyzed the presence of various accessory genes encoding exotoxins and the type of their regulatory gene present, in order to verify the occurrence of some prevalent virulence profile in the types of *agr* found, as well as in the source material of each strain. In addition, an attempt was made to compare the results obtained by various authors with the results found in this study.

To find out whether the isolates analyzed from each center were clonal, molecular typing of these isolates was carried out using the Pulsed Field Gel Electrophoresis (PFGE) technique. According to TERNOVER et al. (1995), molecular typing analysis should be applied to isolates that are related to at least one of the following: (i) a specific area in which the infection is occurring, (ii) a common period for a given infection, (iii) or a common source of infection. Furthermore, according to the same author, molecular typing studies carried out on isolates which have no epidemiological correlation can lead to erroneous information. Therefore, the molecular typing of the OSSA isolates discussed in this paper was not analyzed as a single set, since they come from both areas and from unrelated periods. This methodology served to highlight the lack of clonal relationship between OSSA isolates from the same location. The profile obtained from pulsed field electrophoresis of the OSSA strains in this study shows that there is no clonal relationship between the strains from the same hospital. Exceptions were isolates C3 and C5, from hospital A, which showed only one difference band, and isolates C31 and C33, from hospital N, which showed two different bands. These exceptions could therefore be related to the same sub-clone, according to the criteria presented by TERNOVER et al (1995).

According to PEACKOC et al. (2002) the number of virulence-associated genes carried by a bacterial strain is a product of the interaction between gene acquisition rates, the cost of biological maintenance and the failure rate of the disease-causing strain. As the vast majority of serious *Staphylococcus aureus* infections cannot be explained by the action of just one particular virulence factor, the action of several of these factors during the infection process is imperative.

The survival and proliferation of a given pathogen within a host are favored by mechanisms that allow the evasion of the host's defense mechanisms, as well as the colonization of the microorganism in this microenvironment. Consequently, strains that more efficiently evade the host's defense mechanisms will be the most prevalent types found in such a microenvironment.

Statistical analysis of a large database of pathogenic strains has provided important information regarding genes linked to the virulence of a given prevalent group of these strains. There is evidence that some clonal types are more virulent than others and that these appear more frequently among isolates from patients than in healthy carriers (MELLES et al., 2004).

In previous studies, staphylococcal exotoxins have been associated with specific staphylococcal syndromes. Thus, there have been associations between TSST-1 and staphylococcal toxic shock syndrome (TODD et al., 1985), the epidermolysins ETA and ETB with SSSS (PIEMONT et al., 1988), Panton-Valentine leukocidin (PVL) with the occurrence of boils (PREVÓST et al., 1995) and the presence of SEA-producing strains with post-antibiotic staphylococcal diarrhea (GRAVET et al., 1999). Previous studies have also associated the number of exotoxin genes with the virulence potential of certain staphylococcal strains. Thus, YAGI et al. (2004), studying strains of *Staphylococcus aureus* isolated from patients with atopic dermatitis, when examining the toxin genes of strains isolated from different anatomical sites of these patients, observed that most of the toxin-producing strains (SEA- SED, TSST-1, ETA and ETB) corresponded to strains from the area of exudative lesions (75.4%). However, only 40.7% of the strains isolated from areas without lesions were producers of these exotoxins.

The ability of *S. aureus* strains to evade the host's defense system is attributed, at least in part, to their ability to produce enterotoxins (LE LOIR et al., 2003). Enterotoxin-producing strains can be isolated from various materials, whether clinical or not. The literature shows very variable results regarding the occurrence of enterotoxigenic strains of *Staphylococcus aureus* isolated from both clinical and food samples (ROSEC et al., 1997; GRAVET et al., 2001). This is probably due to differences between the types of materials examined, number of samples, detection methods used, as well as the geographical origin of the strain. Among *S. aureus* strains isolated from food, the percentage of enterotoxigenic strains is estimated at around 25%. However, these estimates vary depending on the type of food and the location of the study (LE LOIR et al.; 2000). In France, among

61 strains isolated from food, 15.9% were enterotoxigenic (ROSEC et al.; 1997). In the Slovak Republic, out of 51 strains of *Staphylococcus aureus* isolated from food, 39.2% were positive for enterotoxins (SEA - 5.9%; SEB - 23.5; SEA and SEB - 9.8%).

In addition to their enterotoxigenic action, *S. aureus* enterotoxins have also been associated with other clinical manifestations. In this regard, GRAVET et al. (2001) observed that SEB was produced in 52% of isolates from patients with impetigo and in 32% of isolates from patients with boils. MULLARY (2001), studying 255 strains of *S. aureus* isolated from bovine mastitis, concluded that the most common genotype of *S. aureus* found was the type against which neutrophils and, above all, the host's defense mechanisms were less efficient. This suggests that the types found in high prevalence have unique characteristics that help them evade the host's immune system, since there are data confirming the activity of *S. aureus* enterotoxins in protecting against neutrophil-mediated apoptosis (MOULDING et al., 1999). However, there are carriers of *S. aureus* strains that carry genes for enterotoxins and do not show any clinical manifestations. In a Spanish study, 269 isolates collected from the nasal microbiota of healthy carriers and from food were analyzed, and positivity for enterotoxins was obtained in 23.9% of the strains from carriers and 26% of the strains from food (LE LOIR et al.; 2003). In addition, the frequency of the type of enterotoxin found is variable, but there is a predominance of some of them. The literature shows that SEA is the type of enterotoxin most commonly found in outbreaks of staphylococcal food poisoning. LE LOIR et al. (2003), analyzing 269 isolates collected from the nasal microbiota of healthy carriers and from food, found that SEA and SEC were the most frequent in the samples analyzed (9.7% for both), and the least frequent was SEB (3%). This author also concluded that SEA and SEC were related to organisms isolated from outbreaks of staphylococcal food poisoning. This is important since there is evidence of transmission from carriers of these strains to food. MULLARY (2001), studying 255 strains of *S. aureus* isolated from bovine mastitis, found that, among the isolates positive for SEs, 62.4% were positive for SEA, 5.4% for SEC,

3.2% for SEB and 3.2% for SED. YAGUI et al. (2004) observed that, of the 81 patients studied with atopic dermatitis, 44% were positive for SEB, 27.2% for SEA, 27.2% for SEC and 4% for SED. In this study, the most detected exotoxin gene corresponded to *seb*. On the other hand, SEB is suggested to be the mediator that induces the lesion of atopic dermatitis as a superantigen by these authors, although they also mention the importance of the presence of ETs in this clinical manifestation. It can thus be seen that the frequency of certain types of enterotoxin depends on the clinical condition of the host.

In the current study, 54.1% of the strains analyzed were positive for at least one enterotoxin gene. For the samples from food related to food poisoning (n=7), 100% of the strains analyzed were positive for at least one enterotoxin gene, serving as a positive control, since their expression had already been previously identified by phenotypic tests, such as indirect hemagglutination, for the determination of enterotoxins.

Strains with enterotoxin-encoding genes are also found in animals, according to recent studies. In a study carried out in Minas Gerais, Brazil, a 43% positivity rate for enterotoxins (SEA - SEE) was found in *S. aureus* strains isolated from bovine mastitis (CARDOSO et al.; 1999). In Denmark, of the 414 strains of Staphylococcus aureus isolated from bovine mastitis, only one carried the gene for SEs (LARSEN et al.; 2000). In Germany, 72.8% of S. aureus isolates from bovine mastitis were enterotoxigenic (SEA-SEJ) (AKINEDEN et al., 2001).

This study found that the most frequent enterotoxin gene in the 37 strains analyzed was the *sea* gene (40.5%), followed by *sec* (34.3%), *seb* (18.9%) and *sed* (5.5%). On the other hand, the *see* gene was not detected in any strain. With regard to the materials analyzed, the largest universe was represented by isolates from blood and food. Although the presence of the SEB toxin in food contaminated by staphylococcal enterotoxins was highlighted by YAGUI et al., 2004, in this study no *S.aureus* sample isolated from food presented the *seb* gene. With regard to the association between enterotoxin genes and the type of their *agr* regulator, there was a 100% association between the presence of the *sea* gene and type III *agr*. Of all the strains carrying one of the enterotoxin genes studied, only one strain, which was positive for *seb*, was classified as type *agrI*. In this way, it can be concluded that the strains with enterotoxin genes preferentially fell into agr types II and III, revealing a characteristic that may be inherent to certain countries, since JARROUD et al. (2001) observed that the strains causing enterotoxin-mediated diseases in France belonged mainly to *agr* groups I and II.

The *tst* gene encodes the production of TSST-1, which is one of the staphylococcal exoproteins responsible for TSS. TSST-1 is produced in around 100% of *S. aureus* strains isolated from the cervical or vaginal mucosa of women with SST, and in approximately 50% of isolates from other body sites of patients with non-menstrual SST. For reasons that are not fully known, certain DNA segments encoding staphylococcal SAGs do not coexist in the same strain. For example, SEB and TSST-1 are rarely produced by the same clinical isolate. It is speculated that in strains carrying certain genetic segments (e.g. SaPI1), the addition region for a second pathogenicity island (e.g. SaPI2 or 3) is modified in such a way as to prevent its integration into the chromosome. Further research should elucidate the mechanisms by which these genetic elements can exclude the integration of these "competing" elements (MCCORMICK et al, 2001).

Of the isolates analyzed in this study, 25% were positive for *tst*. All the isolates from both pleural fluid and cerebrospinal fluid had the *tst* gene. Most of the isolates carrying the *tst* gene (66.7%) were classified as *agr* type III, with only one belonging to *agr* type I.

Isolates carrying the *tst* gene can be found in both hospital and community settings. Deuremberg et al. (2005) found 24% of 51 community *Staphylococcus aureus* positive for the *tst* gene, while 14% of 36 hospital *S. aureus* strains were positive for *tst*. On the other hand, these authors found the presence of *tst* in 25% of *S. aureus* isolates from patients with *Wegener's granulomatosis*, a disease in which the patient is weakened by inflammation of the blood vessels, stipulating a relationship between this disease and the presence of such strains. YAGUI et al. (2004) found that 16.0% of the 81 patients studied with atopic dermatitis were positive for TSST-1. JARROUD et al. (2001) reported that, in their study, the *tst* gene was *not* found in the same strain carrying *LukE-LukD*.

Strains carrying the *tst* gene have also been found in animals, as reported by MULLARY (2001), studying 255 strains of S. aureus isolated from bovine mastitis, where 4.3% of these strains carried the gene for *tst*.

Staphylococcus aureus strains produce two classes of **bicomponent toxins**, composed of two proteins with synergistic actions, secreted independently, referred to as class F and class S components. Four *loci* encoding bicomponent leukotoxins have been characterized in *S. aureus*: *pvl*, *hlg*, *LukE-LukD* and *lukM-LukF'-PV*. This study found the presence of *the pvl*, *hlg* and *lukE-lukD* genes, which are frequently cited in the literature as representatives of the genes encoding the bi-component toxins. Of all the strains analyzed, 97.3% were carriers of the *hlg* gene, within the published statistics. The frequency of the *lukE-lukD* gene was also high, with 96.15% of these strains carrying *it*.

With regard to the presence of the *pvl* gene, the frequency found was 19.4%. This frequency is low compared to some published studies (GRAVET et al., 2001; VON EIFF et al., 2004), which report a high prevalence of this gene in strains isolated from patients with cutaneous manifestations, which does not fit in with our study group, as most of them were isolated from biological liquids. However, this prevalence is high when compared to the stipulated frequency of 2% for clinical strains presenting this gene (PREVÓST et al., 1995). Thus, the relatively high frequency of this gene may reflect the high virulence potential of these strains. Of the seven strains that tested positive for the *pvl* gene, five (71.4%) fell into the agrIII type, and the other two belonged to agr I and II types.

The frequent isolation of leucotoxin-producing strains of *Staphylococcus aureus* in patients with necrotic epidermal lesions, furuncles and osteomyelitis suggests that these toxins play an important role in staphylococcal virulence, at least with regard to skin and tissue infections (VON EIFF et al., 2004). This is evidenced in isolates of *Staphylococcus aureus* carrying the *pvl* gene from immunocompetent patients, who presented with a highly lethal type of necrotizing hemorrhagic pneumonia (GILLET et al., 2002).

The *pvl* gene is found in around 2% of clinical strains, and is strongly associated with the occurrence of boils, while the genes coding for gamma hemolysins are found in around 99% of clinical isolates of *S. aureus* (PREVÓST et al., 1995). GRAVET et al. (2001) observed that all strains of *Staphylococcus aureus* isolated from boils were positive for *pvl*. In addition, nasal isolates from carriers of this gene can serve as an endogenous reservoir for disseminated clinical infections and can spread to other patients (VON EIFF et al., 2004). Consequently, one strategy for reducing hospital infection with these strains would be to eradicate them from nasal carriers (ARCHER et al., 2001).

The LukE-LukD toxin has been shown to be highly frequent (82%) in *Staphylococcus aureus* strains carrying this gene (VON EIFF et al., 2004), with the highest rate of isolates among blood isolates.

In *Staphylococcus aureus* strains isolated from patients with bullous impetigo, the association of this clinical manifestation with the presence of LukE-Luk-D leucotoxins was demonstrated in 78% of the cases studied (GRAVET et al., 2001). GRAVET et al. (2001) in a study in Guyana, showed an association between strains of *Staphylococcus aureus* isolated from boils with the presence of the *pvl* gene; the same was reported for isolates of *Staphylococcus aureus* in Africa (BABA et al., 1999).

With regard to the drastic biological consequences secondary to damage to cell membranes caused by leukocidins, which contribute directly or indirectly to the pathogenesis of acute or chronic disease, data on the molecular epidemiology of these toxins is still limited and restricted, especially with regard to strains *of Staphylococcus aureus* isolated in Brazil.

There is a description of two serologically distinct exfoliative toxins, designated as exfoliative **toxin A** and **exfoliative toxin B** (**ETA** and **ETB**, respectively), whose genes are designated as *etA* and *etB.* Five percent of clinical isolates of *Staphylococcus aureus* produce ETs (PIEMONT et al., 1984), and there are reports of geographical differences in the frequency of expression of these toxins (LISA, 2004). In the present study, no strain was found to carry the *eta* and *etb* genes. This fact is justified by the strong association of ETs with some specific syndromes (SSSS and bullous impetigo), the

which were not within the universe of clinical material analyzed. Another important fact was that no type of *agr* IV was found in the strains analyzed, a type that is strongly related to TEs.

In North America, Europe and Africa, where less than 5% of clinical isolates of *Staphylococcus aureus* produce TEs, ETA is the predominant TE (LADHANI, 2001). In contrast, ETB was the most common reported in Japan (KONDO et al., 1975). The frequency of genes for ETs in isolated strains of *Staphylococcus aureus* varies according to specific clinical manifestations. On the other hand, it is known that the incidence of these strains in healthy patients is less frequent (ZOLLNER et al., 2000).

DANCER & NOBILE (1991) reported that 3% of *S. aureus* strains isolated from the nasal cavity, axilla or perineum of pregnant women were carriers of ET-producing strains. YAGUI et al. (2004), in patients with atopic dermatitis, found that 39.9% of the 81 patients studied were positive for ETA, while 34.6% of the patients were positive for ETB. YAMASAKI et al. (2005) reported that in 103 children with generalized SSSS, the *S. aureus* strains isolated had at least one *eta* or *etb* gene; 30% were infected with isolates carrying only *eta*, 19% carried only *etb* and 51% carried both genes. GRAVET et al. (2001), in a study in Guyana, showed that isolates of *Staphylococcus aureus* from patients with impetigo produced ETs in 86%". In addition, a strong association between TEs and *agr* type IV has been reported (JARRAUD et al., 2000).

In this study, in addition to the gamma-hemolysin already described along with the bi-component toxins, the presence of alpha- and beta-hemolysins was found. These hemolysins are frequently found in strains of *Staphylococcus aureus*. ARESTRUP et al. (1999) found that 56% of *Staphylococcus aureus* strains isolated from healthy patients carried the beta-hemolysin gene and 57% of those isolated from patients with bacteremia carried this gene. The gene for alpha-hemolysin was found in all the strains analyzed.

In our work, 100% of the strains analyzed had the alpha-hemolysin gene, in line with work by ARESTRUP et al. (1999). The gene coding for beta-hemolysin was found to be positive in 78.4% of the strains analyzed, and specifically in the strains isolated from food, there was a high positivity for this gene (85.7%).

Considering the multiple forms and sites of staphylococcal infections, it is natural that these bacteria are well equipped with sensors that indicate the conditions of their microenvironment and regulate the expression of virulence factors, enabling them to colonize, hinder immune defences, multiply and disperse throughout the body (GRAVET et al., 2001). The expression of most virulence factors in *Staphylococcus aureus* is controlled by the *agr* locus, which encodes a cascade regulated by two components, whose activating ligand is a bacterial density-sensing peptide (self-inducing peptide), also encoded by *agr* (NOVICK et al.,2000). Based on the amino acid sequence polymorphism of the autoinducer peptide and its corresponding receptor (AgrC), *Staphylococcus aureus* strains can be divided into four main groups. Associations between a particular type of *agr and* a specific staphylococcal syndrome have been demonstrated for Toxic Shock Syndrome (TST) and Scalded Skin Syndrome (SSSS). In previous studies, it was observed that **TSST-1-producing** isolates fell mainly into the ***agr* III** group (JI et al., 1997). **Exfoliatin-producing** strains, responsible for SSSS, belonged to ***agr*** group **IV** (JARRAUD et al.; 2000). In a study covering 192 strains of *Staphylococcus aureus* isolated from carriers, the prevalence of the ***agrI*** group was observed**,** the majority of which were **resistant to methicillin** (VAN LEEUWEN et al., 2000).

JARROUD et al. (2001) observed that the strains causing **enterotoxin-mediated diseases** belonged mainly to ***agr*** groups **I and II and III, and** that the distribution of *agr* types in the strains studied were: I (43.6%), II (28.0%), III (24.7%) IV (3.3%). MOORE et al. (2001), studying MSSA, found the following prevalence of *agr* types: I (61%), II (26%) and III (13%) and also reported the association of ***agr*** type **III** with ***tst***, not finding this toxin together with *seb* in the same strain.

In our study, most of the OSSA isolates analyzed fell into *agr* type III (62.2%). The rest were equally distributed between *agr* types I and II (18.9% each). It can thus be seen that the oxacillin-sensitive *Staphylococcus aureus* strains isolated in Brazil used in this study, due to the predominance of agr type III (62.2%), differ from studies published in other countries, such as the predominance of *agr* type I (43.6%) in *Staphylococcus aureus* strains observed in France (JARROUD et al.; 2001) and England (MOORE et al., 2001).

With regard to the type of clinical material, the two isolates from pleural fluid fell into *agr* type I, while the two isolates from cerebrospinal fluid fell into *agr* type III. This type of *agr* corresponded to the majority of strains isolated from blood cultures (57.7%) and food (71.4%). Therefore, *agr* type III proved to be the most frequent among the isolates from the different clinical materials analyzed, with the exception of that observed in the strains from pleural fluid.

Curiously, 100% of the strains carrying the *sea* gene belonged to the *agr* III type, drawing special attention to this association. The same predominance of the agr III type was observed for the other genes analyzed: pvl (71.4%), lukE-lukD (62.9%), hla (62.8%), hlb (68.9%), hlg(61.2%).

According to MOORE et al. (2001), the *tst* gene belonged mostly to *agr* type III (66.7%).

It was therefore found that *Staphylococcus aureus* can use combinations of various virulence factors, which can result in a clinical syndrome and possibly contribute to bacterial dispersal in the host's organism. Certain accessory genes, coding for exoproteins, can be used as potential virulence markers for such strains, such as the *seb, sea, tst* and *pvl* genes, which were found with relevant frequency in these studies. In addition, agr type III can be considered a potential representative of successful OSSA strains in the hospital or community environment.

5. Conclusion

This study found that the frequency of exotoxin-encoding genes varies due to various factors, such as the nature of the strain's isolation source and geographical location. Thus, it was observed that the most frequently determined enterotoxin-encoding gene was *sea*, followed by *sec*, and that *sea was* 100% related to the *agr* III type. As there was a low frequency of isolates carrying *seb*, further research is suggested to check the frequency of *S. aureus* isolates from patients with atopic dermatitis with the presence of these genes, as this correlation was found in the literature.

The results diverged from those of studies carried out in other countries, such as the finding of strains containing genes coding for *seb* and *tst* (33%), a correlation that the researched literature thought was not frequent, as well as the simultaneous presence of *tst* and *lukE-lukD*.

No isolates carrying exfoliative toxins were found in the universe studied. It is therefore important to check for the presence of these genes in strains from patients with epidermal manifestations such as SSSS and bullous impetigo. It is also important to check the relationship between strains that may have these genes and agr type IV, which was not found in the isolates analyzed.

Despite the low number of isolates from CSF (n=2) and Pleural Fluid (n=2), it was observed that all these isolates were positive for *pvl*, indicating that, in invasive infections, the participation of Panton-valentine type leucocidins is possible. The relatively high frequency (14%) of isolates carrying this gene found in this study highlights the participation of the PVL leukocidin in increasing the virulence of these isoalds.

An interesting feature of oxacillin-sensitive *S. aureus* isolates is that they can present a significant number of exotoxin-encoding genes. Depending on the presence and expression of these genes, various clinical manifestations can be presented by the host. It is therefore extremely important to know the regulatory genes that control these gene expressions. In this respect, it was found that the majority (62%) of the isolates analyzed were carriers of the *agr* III type, diverging from results published in other countries which report a predominance of *agr* I.

It can therefore be concluded that the molecular basis of the pathogenicity *of S. aureus is* multifactorial, depending not only on the presence but also on the expression of various accessory genes. Thus, it can be seen that the serious infections triggered by these bacteria are the product of the sum of several virulence factors, encoded by such genes. In this context, it is extremely important to know the genotype profile of *S. aureus* strains that are frequently isolated from a given location, so that we can not only predict clinical manifestations resulting from possible dispersion, but also find mechanisms to help control them.

6. References

AMAGAI, M.; YAMAGUCHI, T.; HANAKAWA, Y.; NISHIFUGI, K.; SUGAI, M.; STANLEY, J.R. Staphylococcal exfoliative toxin B specifically cleaves desmoglein 1. **J. Invest. Dermatol.** V. 118, p. 845-850, 2002.

AKINEDEN, °; ANEMULLER, C.; HASSAN, A.A.; LAMMLER, C.; WOLTER, W.; ZSCHOC, M. Toxin genes and other characteristics of *Staphylococcus aureus* isolates from milk of cows with mastitis. **Clin. Diagn. Lab. immunol**. v. 8, p. 959964,2001.

ARCHER, G.L. Staphylococcus aureus: a well-armed pathogen. **Clin Infect Dis.** v. 26,p. 1179-1181, 1998.

ARCHER, G.L.; CLIMO, M.W.; *Staphylococcus aureus* bacteremia - consider the source. **N Engl J Med.** V.344, p. 55-56, 2001.

ARESTRUP, F.M.; LARSEN, H.D.; ERIKSEN, N.H.; ELSBERG, C.S.; JENSEN, N.E. Frequency of alpha- and beta haemolysin in *Staphylococcus aureus* of bovine and human origin. A comparison between phenol- and genotype and variation in phenotypic expression. **APMIS.** V.107, p.425-430, 1999.

BALABAN, N.; GOLDKORN, T.; GOV, T.; HIRSHBERG, M.; KOYFMAN, N.;

MATHEWS, H.R.; NHAN, R.T.; SINGH, B.; UZIEL, O. regulation of *Staphylococcus aureus* pathogenesis via target of RNA-III-activating protein (TRAP). **J Biol Chem**. v. 276, p. 2658-2667, 2001.

BERGDOLL, M.S.; CRASS, B.A.; REISER, R.F.; ROBBINS, N.R.; DAVIS, J.P. A new staphylococcal enterotoxin, enterotoxin F, associated with the Toxic Shock syndrome *Staphylococcus aureus* isolates. **Lancet j.** v.8228, 1017-1027, 1981.

BHAKDI, S.; TRANUM, J.J. Alpha-toxin of *Staphylococcus aureus*. **Microbiol.Review**. v. 55, p.733, 1991.

BHAKDI, S.; MUHLY, M.; MANNHARDT, U.; et al. Staphylococcal alpha-toxin promotes blood coagulation via attack on human platelets. **J Exp Med**. v.168, p.527, 1988.

BERGDOLL, M.S.; CRASS, B.A.; REISER, R.F. et al. A new staphylococcal enterotoxin, enterotoxin F, associated with Toxic Shock Syndrome *Staphylococcus aureus* isolates. **Lancet**. v.1, p1017, 1981.

BERNHEIMER, A.W.; RUDY, B.; Interactions between membranes and cytolytic peptides. **Biochim Biophys Acta.** v.864, p.123, 1986.

BETTLEY, M.J.; BORST, D.W.; REGASSA, L.B. Staphylococcal enterotoxins, toxic shock syndrome toxin and streptococcal pyrogenic exotoxins: A comparative study of their molecular biology. **Chem.**

Immunol. v. 55, p. 1-35, 1992

BETLEY, M.J.; MEKALANOS, J.J. Staphylococcal enterotoxin A is encoded by phago. **Science**. v. 229, p. 185-187, 1985.

BOHACH, G.A.; FAST, D.J.; NELSON, R.D.;SCHLIEVERT, P.M. Staphylococcal and Streptococcal pyrogenic toxins involved in toxic shock syndrome and related illnesses. **Crit Rev Microbiol**. v. 17,p. 251-272,1990

BOHACH, G.A.; FOSTER, T.J. Gram-Positive Pathogens. **American Society for Microbiology**. p. 367-378, 2000.

BOHACH, G.A.; FAST, D.J. NELSON, RD.;SCHLIEVERT, P.M. Staphylococcal and Streptococcal pyrogenic toxins involved in toxic shock syndrome and related illnesses. **CRC Crit Rev Microbiol**. v.17, p.251, 1990.

BOYCE, J.M.; The staphylococci in human disease. **Epidemiology and prevention of nosocomial infections**. Churchill Livingstone, New York, N.Y.. p. 309-329, 1997.

BRMLEY, A.J.; PATEL, H.A.; REILLY, M.O. et al. Roles of alpha-toxin and betatoxin in virulence of *Staphylococcus aureus* for the mouse mammary gland. **Infect Immun**. v. 57, p. 2489, 1989.

CALLEGAN M.C.; ENGEL, L.S.; HILL, J.M.; O'CALLAGHAN, R.J. Corneal virulence of *Staphylococcus aureus*: roles of alpha-toxin and protein A in pathogenesis. **Infect. Immun**. v.62, p.2478, 1994.

CARDOSO, H.F.; SILVA, N.; SENA,M.J.; CARMO, L.S. Production of enterotoxins and toxic shock syndrome toxin by *staphylococcus aureus* isolated from bovine mastitis in Brazil. **Lett. Appl. Microbiol**. v.29, p. 347-349.

COONEY, J.; KIENLE Z.; FOSTER, T.J., O'TOOLE, P.W. The gamma hemolysin locus of *Staphylococcus aureus* comprises three linked genes, two of which are identical to the genes forming F and S components of leukocidin. **Infect Immun**. v.61, p.768, 1993.

CORKILL, E.J.;ANSON, J.J.; GRIFFTS, P.; HART, C.A. Detection of elements of the Staphylococcal cassette chromosome in a methicillin susceptible (*mec* A gene negative) homologue of a fucidin-resistant MRSA. **Journal of Antimicrobial Chemotherapy**. v.54, p. 229-231,2004.

DANCER S.J.; NOBLE, W.C. Nasal, axillary, and perineal carriage of *Staphylococcus aureus* among women: identification of strains producing epidermolytic toxin. **J. Clin. Pathol.** V.44, p.681-684, 1991.

DAY, N.P.J.; MOORE, C.E.; ENRIGHT, M.C.; BERENDT, A.R., SMITH, J.M.; MURPHY, M.F.; PEACOCK, S.J.; SPRATT, P.G.; FEIL, E.J. A link between virulence and ecological abudance in natural populations of Staphylococcus aureus. **Science**. v. 292,p. 114-116, 2001.

DEURENBERG, R.H.; NIEUWENMHUIS, R.F.; DRIESSEN, C.; LONDON, N.; STASSEN, F.R.; VAN

TIEL, F.H.; et al. The prevalence of the *Staphylococcus aureus* tst gene among community- and hospital- acquired strains and isolates

from Wegener's Granulomatosis patients. **FEMS Microbiol. Lett**. V.245, p. 185189, 2005.

DINGES, M.M.; ORWIN, P.M.; SCHLIEVERT, P.M. Exotoxins of *Staphylococcus aureus*. **Clin. Microbiol**. v. 13, p.16-34, 2000.

FiNK, D.; CONTRERAS, M.L.; LELKES,P.I.; LAZAROVICI,P. Staphylococcus aureus alpha-toxin activates phospholipases and induces a Ca^{2+} influx in PC12 cells. **Cell. Signal**. V.1(4), P. 387-393 ,1989.

FTZGERALD, J.R.; MONDAY, S.R.; FOSTER, T.J.; BOLACH; G.[a]; HARTIGAN, P.J.; et al. Characterization of putative pathogenicity island from bovine

Staphylococcus aureus encoding multiple superantigens. **J. Bacteriol**. v.183, p6370, 2001.

FITZGERALD, j.R.; STURDEVANT, E.D.; MACKIE, S.M.;GILL, S.R.; MUSSER, j.M. Evolutionary genomics of *Staphylococcus aureus:* Insights into the origin of methicillin-resistant strains and the toxic shock syndrome epidemic. **PNAS.** V.98, p.8821-8826, 2001.

GEMMEL, C.G. Staphylococcal scalded skin syndrome. j. Med. Microbiol. V. 43, p. 318-327, 1995.

GILLET, Y.; ISSARTEL, B.; FOURNET, J.C.; LINA, G.; BES, M.; et al. association between *Staphylococcus aureus* strains carrying gene for Panton-Valentine leukocidin and highly lethal necrotizing pneumonia in young immunocompetent patients. **Lancet.** V.359, p. 753-759, 2002.

GOTTILIEB, G.S.; FOWLER, V.G.; KANG, L.K. *Staphycoccus* bacteraemia in the surgical patient: a prospective analysis of 73 postoperative patients who developed *Staphylococcus aureus* bacteraemia at a tertiary care facility. **J Am Coll Surg**. v. 190,p. 50-57, 2000.

GRAVET, A.; COLIN, D.A.; KELLER, D.; GIRADOT, R.; MONTEIL, H.; PREVÓST, G. Characterization of a novel structural member, LukE-LukD, of the bi-component staphylococcal leukotoxins family. **FEBS Lett.** V. 436, p. 202-208, 1998.

GRAVET, A.; RONDEAU, M.; MONTÉIL, C.H.; GRUNEMBERGER, F.; MONTEIL, H.; et al. Predominat *Staphylococcus aureus* isolated from antibiotic-associated diarrhea is clinically relevant an produces enterotoxin Aand the bicomponent toxin LukE-LukD. **J. Clin. Microbiol**. V. 37, p. 4012-4019, 1999.

GRAVET, A.; COUPPIE, P.; MEUNIER, O.; et al.; *Staphylococcus aureus* isolated in cases of impetigo produces both epidermolisin A or B and LukE-Lukd in 78% of 131 retrospective and prospective cases. **J. Clin. Microbiol**. V. 39, p. 4349-4356, 2001.

HARRIS, T.O.; BETLEY, M.J.; Biological activities of staphylococcal enterotoxin type A mutants with N-terminal substitutions. **Infect Immun**. v.63, p.2133, 1995.

HARRIS, T.O.; GROSSMAN, D.; KAPPLER, J. W.; MARRACK, P.; RICK, R.R.;

BETLEY, M.J.; et al.;. Lack of complete correlation emetic and T-cell-stimulatory activities of staphylococcal enterotoxins. **Infect. Immun**. v. 61, p. 3175-3183, 1993.

HELALI, N.E.; CARBONNE, A.; NAAS, T.; KERNEIS, S.; FRESCO, O.; GIOVANGRANDI, Y.; et al. Nosocomial outbreak of staphylococal scalded skin syndrome in neonates: Epidemiological, investigation and control. **J. Hosp. Infect.** V. 20, p.1-9, 20005.

HERNANDEZ, F.J.; GOYACHE, J.; ORDEM, J.A.; BLANCO, J.L.; DOMENECH, [a];

SUAREZ, G. et al. Repair and enterotoxin synthesis by Staphylococcus aureus after tthermal shock. **Appl. Environ. Microbiol**. v. 59, p. 1515-1519, 1993.

HERZER, C.M. Toxic shock syndrome: Broadening the differential diagnosis. **J. Am Board Fam Pract.** V. 14, p. 131-136, 2001.

HOLEKOVA, B.; HOLODA, E.; FOTTA, M.; KALINASKOVA, V.; GONDOL, J.;

GROLMUS, J. Occurrence of enterotoxigenic *Staphylococcus aureus* in foood. **Ann Agric Environ**. Med. v.9 p. 179-182, 2002.

HOLMBERG,S.D.; BLAKE P.A. Staphylococcal food poisoning in the United States. New facts and old misconceptions. **JAMA**. v.251, p.487, 1984.

JARRAUD, S.; LYON, G. J.; FIGUEIREDO, A.M.; LINA,G.; VANDENESCH, F.;

ETIENE, J.; MUIR,T.W.; NOVICK, R.P. Exfoliatin-producing strains define a fourth *agr* specificity group in *Staphylococcus aureus.* **J. Bacteriol**. v.182, p.6517-6522, 2000.

JARRAUD, S.; PEYRAT, M.A.; LIM, A.; TRISTAN, A.; BES, M.; et al. A high prevalent operon of enterotoxin gene, forms a putative nursery of superantigens in *Staphylococcus aureus*. **J. Immun**. v. 166, p. 669-677, 2001.

Jarraud, S.; Mougel, C.; Thioulouse, J.; Lina G.,[1] Meugnier, H.; Forey F.; Nesme X.; Etienne, J.; Vandenesch, F. Relationships between *Staphylococcus aureus* Genetic Background, Virulence Factors, *agr* Groups (Alleles), and Human Disease. **Infect Immun**. v.70, p.631-641,2002.

JI, G.; BEAVIS, R.; NOVICK, R.P.; Bacterial interference caused by autoinducing peptide variants. **Science.** v.276, p.2027-2030, 1930.

JONAS, D.; WALEV, I.; BERGER, T.et al. Novel path to apoptosis: small transmembrane pores created by staphylococcal alpha-toxin in T limphocytes evoke internucleosomal DNA degradetion. **Infect Immun**. v. 62, p.1304,1994.

KANZAKI, H.; UEDA, M.; MORISHITA, Y.; AKIJIAMA, H.; ARATA, A.; KANZAKI, S. Producibility of exfoliative toxin and Staphylococcus coagulase types of Staphylococcus aureus strais isolated from skin infections and atopic dermatitis. **Dermatol.**, V.195, p. 6-9, 1997.

KAPER, J.B.; KACKER, J. Pathogenicity islands and other mobile virulence elements. **ASM Press**.

Washington, DC, 1999.

KONDO, I.; SAKURAI, S.; SARAI, Y.; FUTAKI, S. Two serotypes of exfoliatin and their distribution in staphylococcal strains isolated from patients with scalded skin syndrome. **J. Clin. Microbiol**. V.1, p. 397- 400, 1975.

LADHANI, S.; JONNOU, C.L.; LOCHRIE D.P.; EVANS, R.W.; POSTON, S.M. Clinical, microbial and biochemical aspects of the exfoliative toxins causing

Staphylococcal scalded Skin Syndrome. **Clin. Microbiol. Rev.** v.12, p.224-242, 1999.

LADHANI, S. Recent developments in staphylococcal scalded skin syndrome. **Clin. Microbiol. Infect**. v.7, p. 301-307, 2001.

LAN, R.; REEVES, P.R. Intraspecies variation in bacterial genomes: the need for a species genome concept. **Trends Microbiol**. v. 8,p. 396-401, 2000.

LARSEN, H.D.; HUDA, A.; ERIKSEN, N.H.R.; JENSEN, N.E. Differences between Danish bovine and human *Staphylococcus aureus* isolates in possession of superantigens. **Vet. Microbiol**. v. 76, p. 153-162, 2000.

LE LOIR, I.; BARON, F.; GAUTIER, M. *Staphylococcus aureus* and food poisoning.

Genet. Mol. Res. v.2, p. 63-76, 2003.

LEE, C.Y.; SCHIMIDT, J.J, WINEGAR, J.A.D. et al. Sequence determination of the exfoliative toxin a and toxin b genes from *Staphylococcus aureus*. **J Bacteriol**. v.169, p.3904, 1987.

LINDSAY, J.A.; RUZIN, A.; ROSS, H.F.; KUREPINA, N.; NOVICK, N.P. The gene for toxic shock toxin is carried by a family of mobile pathogenicity islands in *Staphylococcus aureus*. **Mol. microbiol**. v.29, p. 527-543, 1998.

LISA, R.W. *Staphylococcus aureus* exfoliative toxins: How they cause disease. **Derm. Foundat.** V.122, p. 1070-1077, 2004.

MACKENZIE, A.; JOHNSON, W.B.H.; NORRISH, B. JAMIESON, F. A prolonged outbreak of exfoliative toxin A-producing *Staphylococcus aureus* in a Newborn Nursery. **Diagn. Microbiol. Infect. Dis**. V. 21, p.69-75, 1995.

MARR, J.C.; LYON, J.D.; ROBERSOM, J.R. et al. Characterization of novel type c staphylococcal enterotoxins: Biological and evolutionary implications. **Infect Immun**. v.61, p.4254, 1993.

MAYVILLE, P.; JI, G.; BEAVIS, R.; YANG, H.; GOGER,M; NOVICK, R.P.; MUIR, T.W. Structure-activity analysis of synthetic autoinducing thiolactone peptides from *Staphylococcus aureus* responsible for virulence. **Proc. Natl Acad Sci USA** v.96, p.1218-1223, 1999.

MARRACK, P.; KAPPLER, J. The Staphylococcus enterotoxins and their relatives. **Science.** p. 248-705, 1990.

McCORMICK, J.K; YARWOOD, J.M; SCHLIEVERT, P.M. Toxic shock syndrome and bacterial superantigens: an update. **Annu. Ver. Microbiol**. v.55, p. 77-104, 2001.

MELLES, D. C.; GORKINK, R. F.; BOELENS, H.A.; SNIJDERS, S.V., PEETERS J. K.; MOORHOUSE, M.J.; VAN DER SPEK, P. J.; VAN LEEUWEN, W.B.; SIMONS, G.; VERBRUGH, H.A.; VAN BELKUM, A. Natural population dynamics and expansion of pathogenic clones of Staphylococcus aureus. **J. Clin Invest**. v. 114, p.1732-1740, 2004

MELZER, M.; EYKYN, S.J.; GRANSDEN, W.R.; CHINN S. Is methicillin-resistant *Staphylococcus aureus* more virulent than methicillin-susceptible *S. aureus*? A comparative cohort study of British patients with nosocomial infection and bacteremia. **Clin Infecti Dis.** v. 37,p. 1453-1460, 2003.

MOULDING, D. A.; WATER, C.; HART, C.A.;. EDWARDS, S. W. Effects of Staphylococcal enterotoxins on human neutrophil functions and apoptosis. **Infect. immun.** v.67, p. 2312-2318, 1999.

MOORE, P.C.L.; LINDSAY, J.A. Genetic variation among hospital isolates of methicillin-sensitive *Staphylococcus aureus*: evidence for horizontal transfer of virulence genes. **J Clinl Microbiol**. v. 39, p.2760-2767, 2001

MOUSSA, B.L.; SANNI, A.; DAGNRA, A.Y.; ANAGONOU,S.; DAVID, M.P.; et al.; Approche épidémiologique de l'antibiorésistance et de la proction de leucotoxines par lês souches de *Staphylococcus* aureusisolées en Afrique de l'Ouest. **Med. Mal. Infect**. V. 29, p.689-696, 1999.

MUNSON, S.H.; TREMAINE, L.T.; BETLEY, M.J.; WELK,R.A. Identification and characterization of staphylococcal enterotoxin types G and I from *Staphylococcus aureus*. **Infect. Immun**. v.66, p. 3337-3348, 1998.

MULLARY, I.K.; SU, C.; FRIEZE, N.; PARK, Y.H.; SORDILLO, L.M.

Staphylococcus aureus agr genotypes with enterotoxin production capabilities can resist neutrophil bactericidal activity. v. 69, p.45-51, 2001

NATIONAL COMMITTEE FOR CLINICAL LABORATORY STANDARDS (NCCLS). Performance Standards for Antimicrobial Susceptibility testing. M100- S11. v. 21, 2003.

NOVICK, R.P. Pathogenicity factors and their regulation. In Fischetti,V. A.; Novick, R. P.; Ferretti, J.J.; Portnoy, D.A.; Rood, J.J. Gram-positive pathogens. **ASM, Washington DC.** p 392-407, 2000.

NOVICK, R.P. Autoinduction and signal transduction in the regulation of staphylococcal virulence. **Mol. Microbiol**. V. 48, p. 1429- 1449, 2003.

PARRILO, J.E.: Pathogenetic mechanisms of septic shock. **N. Engl J. M.** v.329, p.1427, 1993

PENG, H.L.; NOVICK, R.P.; KREISWIRTH, B. et. al. Cloning, characterization, and sequence of an accessory gene regulator (agr) in *Staphylococcus aureus*. **J Bacteriol**. v.170, p.4168, 1994.

PIEMONT, Y.; RIFAI, S.; MONTÉIL, H. Les esfoliatines de *Staphylococcus aureus*. **Bull. Inst**. V. 86,

p. 263-296, 1988.

PLOMET, M. Preparation and purification of gamma-hemolysin of staphylococci. **Methods Enzymol**. v.165, p.8, 1988.

PREVOST, G.; JAULHAC B. ; PIEMONT,Y. DNA fingerprinting by pulsed-field gel electrophoresis is more effective than ribotyping in distinguishing among methicillin-resistant Staphylococcus aureus isolates. **J Clin Microbiol.** V. 30(4)p.967-973, 1992.

PREVOST, G.; COUPPIE, P.; PREVOST, P.; GAYET, S.; PETIAU, P. et al. Epidemiological data on *Staphylococcus aureus* strains producing synergohymenotropic toxins. **J Med Microbiol**. v.42 p.237, 1995.

PREVOST, G.; CRIBIER, B.; CUPPIE ,P. PETIAU, P.; SUPERSAC, G.; et al. Panton-valentine leucocidin and gamma-hemolysin from *Staphylococcus aureus* ATCC 49775 are found by distinct genetic loco and have different biological activities. **Infect Immun**. v.63, p.4121, 1995.

PROJAN, S.J.; KORNBLUM,, J.; KREISWIRTH,B. et al. Nucleotide sequence: The beta-hemolysin gene of *Staphylococcus aureus*. **Nucleic Acids Res**. v.17, p.3305, 1989.

RECSEI, P.; KREISWIRTH, B.; O' REILLY, M. et al. regulation of exoprotein gene expression by *agr*. **Mol Gen Genet**. v.202, p.598, 1986.

REINGOLD, A.L.; HARGRETT, N.T., DAN, B.B. et al. Non-menstrual ToxicShock syndrome: A review of 130 cases. **Ann Intern Med**. v.196, p.871, 1982.

RUZICKOVA, V. A rapid method for the differentiation of *Staphylococcus aureus* hemolysins. **Folia Microbiol**. v.39, p.242-308, 1972.

ROSEC, J.P.; GUIRAUD, J.P.; DALET, C.; RICHARD, N. Enterotoxin production by staphylococci isolated from foods in france. **Int. j. food Microbiol**. v. 35, p. 213221, 1997.

SAKO, T.; TSUCHIDA, N. Nucleotide sequence of the staphylokinase gene from *Staphylococcus aureus*. **Nucleic Acids Res**. vol.11, p.7679, 1983.

SAMBROOK, J.; FRITSCH, E.F.; MANIATIS, T. Molecular cloning: A laboratory manual, 2nd ed. **Cold Spring Harbor Laboratory Press, Cold Spring Harbor, N.Y**., 1989.

SATO, H.; MATSUMRI, Y.; TANABLE, T.; SAITO, H.; SHIMIKU, A.; KAWANO,J. A new type of staphylococcal exfoliative toxin from *Staphylococcus aureus* strains isolated from a horse with **phlegmon. Infect. Immun.** V. 62, p-3780-3785, 1994.

SCHLIEVERT, P.M.; SHANDS, K.N.; DAN, B.B. et al. Identification and characterization of an exotoxin from *Staphylococcus aureus* associated with Toxic shock Syndrome. **J Infect Dis** v.143, p.509, 1981.

SCHLIVERT, P.M.;JABLONSKI, L.M.; ROGGIANI, M.; SADLER, I.; CALLANTINE, S.; et al.;

Pirogenic toxin superantigen site specificity in toxic shock syndrome and food poisoning in animals. **Infect. immun.** v. 68, p. 3630-3634, 2000.

SOMERVILLE, G.A.; STEPHEN, B.B.; FITZGERLD, J.R.; DELEO, F.R.; COLE, R.L.; HOLF, J.S.; MUSSER, J.M. In vitro serial passage of *Staphylococcus aureus*: changes in physiology, virulence factor production, and *agr* nucleotide sequence. **J Bacteriol**. v. 184, p. 1430-1437, 2002.

Struelens, M. J.; Deplano, A.; Godard, C.; Maes, N.; Serruys, E. Epidemiologic typing and delineation of genetic relatedness of methicillin-resistant Staphylococcus aureus by macrorestriction analysis of genomic DNA by using pulsed-field gel electrophoresis. **J Clin Microbiol**. v. 30, p. 2599-2605, 1992.

SUTRA,L.; POUTEREL, B. virulence factors involved in the pathogenesis of bovine intramammary infections due to S. aureus. **J. Med. Microbiol**. v.40, p 79-89, 1993.

SUTTORP, N.; SEEGER, W.; DEWEIN, E.; et al. Staphylococcal alpha-toxin stimulates synthesis of prostacylin by cultured endothelial from pig pulmonary arteries. **Am J Physiol**. v.248, p.127, 1985

TENOVER, F.C.; ARBEIT, R.D.; GOERING, R.V.; MICKELSEN,.P.A.;

MURRAY,B.E.; PERSING,D.H.; SWAMINATHAN, B. Interpreting chromosomal DNA restriction patterns produced by pulsed-field gel electrophoresis: criteria for bacterial strains typing. **J. Clin. Mivrobiol.** v 33, p. 2233-2239, 1995.

TODD, J.; FISHAUT, M.;KAPRAL,F.; WELCH,T. Toxic shock syndrome associated with phage-group-I sataphylococci. **Lancet**. v.2, p.1116, 1978.

TODD, J.K. Staphylococcal toxin syndromes. **Annu. Rev. Med.** V. 36, p. 337-347, 1985.

TREMAINE, M.T.; BROCKMAN, D.K.; BETLEY, M.J. Staphylococcal enterotoxin A gene (sea) expression is not affected by the accessory gene regulator (agr). **Infect. Immun**. v. 61, p. 356-359, 1993.

VAN LEUWEN, W.; VAN NIEUWENHUIZEN, C. G.; VERBRUG, H. VAN BELKUN, A. Population studies of methicillin-resistant and sensitive *Staphylococcus aureus* reveal a lack of variability in the *agrD* gene, encoding a Staphylococcal autoinducer peptide. **J. Bacteriol**. v.182, p.5721-5729, 2000.

VATH, G.M.; EARTHART, C.A.; MONIE, D.D.; LANDOLO, J.J.; SCHLIEVERT, P.M.; et al. The crystal structure of exfoliative toxin B: a superantigen with enzymatic activity, Biochemistry, v, 38, p.10239-10246, 1999.

VICENT, J.L.; BIHARI D.J.; SUTER, P.M.; BRUINIG, H.A.; WHITE,J.; WOLF, M.; SPENCER, R.C.; HEMMER,M. The prevalence of nosocomial infection in intensive care units in Europe. **Results of the European Prevalence of Infection in Intensive care (EPIC) Study. EPIC International Advisory Committee. JAMA.** v. 274, p. 639-644, 1995.

YAGUI, S.; WALAKI, N.; IKEDA, N.; TAGAGI, Y.; UCHIDA, H.; KATO, Y.; MINAMINO, M. Presence of staphylococcal exfoliative toxin A in sera of patients with atopic dermatitis. **Clin. Exp. Allerg**. V.

34, p.984-993, 2004

YAMAGUCHI, T.; HAYACHI, T.; TAKAMI, H.; NAKASOME, K.; OHNISH, M.;

NAKAYAMA, K.; YAMADA, s.; KOMATSUZAWA, H.; SUGAI, M. Phage conversion of exfoliative toxin A production in *Staphylococcus aureus*. **Mol Microbiol**. v. 38, p.694-705, 2000.

YAMAGUCHI, T. NISHIFUGI, K.; SASAKI, M. et al. Identification of the *Staphylococcus* aureus etd pathogenicity island encoding a novel exfoliative toxin, ETD, and EDIN-B. **Incfet. Immun**. V.70, p. 5835-5845, 2002.

YAMAZAKI, O.; YAMAGUCHI, T.; SUGAI, M.; CELLIER, C.C.; ARNAUD, F.; VANDENSCH, F.; et al. Clinical manifestations of staphylococcal scalded skin syndrome depend on serotypes of exfoliative toxins. J. Clin. Microbiol. V.43, p.1890- 1893, 2005.

YOSHIZAWA, Y.; SAKURADA, J.; SAKURAI, S.; MACHIDA, K.; KONDO, I.;

MASUDA. S. An exfoliative toxin A-converting phage isolated from *Staphylococcus aureus* strain ZM. **Microbiol Immunol.** v.44, p.189-191, 2000.

WADSTROM, T. MOLLBY, R. Studies on extracellular proteins from *Staphylococcus aureus*. **Biochim Biophys Acta**. v.242, p.308, 1972.

WERNWE, S.; COLIN, D.A.; CORRAIOLA, M; MENESTRINA, G.; MONTEIL, H.; PRÉVOST, G. Retrieving biological activity from LukF-PV mutants combined with different S components implies compatibility between the stem domains of these staphylococcal bicomponent leukotoxins. **Infect. Immun.** V.70 (3), p. 1310-1318, 2002.

ZOLLNER, T.M.; WICHELLAUS, T.A.; HARTUNG, A.; et al. Colonization with superantigen-producing *Staphylococcus aureus* is associated with increased severity of atopic dermatitis. **Clin. Exp. Allerg.** V. 30, p. 994-1000, 2000.

8. Annexes

Annex - List of hospitals/laboratories and their locations

Acronym	Hospital	**City - State**
HP-SP	Public Hospitals	São Paulo -SP
HUSP	University Hospital	São Paulo - SP
HC-SP	Hospital das Clinicas	São Paulo - SP
HU-Stm	University Hospital	Santa Maria - RS
HC-SLUC	Sâo Lucas Hospital	Santa Maria - RS
HC-Rfe	Hospital das Clinicas	Recife - PE
FM-Ub	School of Medicine Hospital Medicine	Uberaba - Minas Gerais
LMFlor	Hospital of Santa Luzia	Florianópolis - SC
HUBel	University Hospital	Belém - Parà
HUMnaus	*University Hospital*	*Manaus - Amazonia*
HUVit	University Hospital	Vitória -ES
HUBsl	University Hospital	Brasilia - DF
LARP	Laboratory of Food	Riberâo Preto - SP
HG	Military Hospital	São Paulo

Annex 2 - List of PCR results obtained for the genes analyzed in oxacillin-sensitive strains

Isolated	Proc.	F. Isol.	*agr*	*sea*	*seb*	*sec*	sed	*see*	*tst*	*eta*	*etb*	*pvl*	*lucd*	*hla*	*hlb*	*hlg*
C1	A	LCR	III	+	-	+	-	-	+	-	-	-	+	+	-	+
C2	A	LP	I	-	+	-	-	-	+	-	-	+	+	+	+	+
C3	A	Sg	III	+	-	+	-	-	-	-	-	+	-	+	+	+
C4	A	LCR	III	-	+	-	+	-	+	-	-	+	+	+	+	+
C5	A	LP	III	+	-	+	-	-	+	-	-	-	-	+	+	+
C6	B	*Sg*	III	-	-	-	-	-	-	-	-	-	+	+	-	-
C7	B	Sg	III	+	-	+	-	-	-	-	-	-	+	+	+	+
C8	C	Sg	I	-	-	-	-	-	-	-	-	-	+	+	+	+
C9	C	Sg	II	-	-	-	-	-	-	-	-	-	+	+	-	+
C10	C	Sg	I	-	-	-	-	-	-	-	-	-	+	+	-	+
C11	D	Sg	II	-	-	-	-	-	-	-	-	-	+	+	-	+
C12	D	Sg	III	-	-	-	-	-	-	-	-	-	+	+	+	+
C13	E	Sg	III	-	-	-	-	-	-	-	-	-	+	+	+	+
C14	E	Sg	II	-	+	-	+	-	-	-	-	-	+	+	+	+
C15	F	Sg	III	-	+	-	-	-	+	-	-	-	+	+	+	+
C16	F	Sg	III	-	+	-	-	-	-	-	-	-	+	+	+	+
C17	G	Sg	III	+	-	+	-	-	-	-	-	-	+	+	+	+
C18	H	*Sg*	III	-	-	-	-	-	-	-	-	-	+	+	+	+
C19	H	Sg	III	-	-	-	-	-	-	-	-	-	+	+	+	+
C20	H	Sg	III	-	+	-	-	-	-	-	-	-	+	+	+	+
C21	H	Sg	III	-	+	-	-	-	-	-	-	-	+	+	+	+
C22	I	Sg	III	-	-	-	-	-	-	-	-	-	+	+	+	+
C23	J	Sg	III	-	-	+	-	-	-	-	-	-	+	+	+	+
C24	L	Sg	II	-	-	+	-	-	-	-	-	-	+	+	+	+
C25	L	Sg	I	-	-	-	-	-	-	-	-	-	+	+	+	+
C26	L	Sg	III	+	-	+	-	-	-	-	-	+	+	+	+	+
C27	L	Sg	III	+	-	+	-	-	-	-	-	+	+	+	+	+
C28	M	Sg	II	-	-	-	-	-	-	-	-	+	+	+	+	+
C29	N	Sg	I	-	-	-	-	-	+	-	-	-	+	+	-	+
C30	N	Sg	II	-	-	-	-	-	+	-	-	-	+	+	-	+
Al-1	O	Al	III	+	-	+	-	-	+	-	-	-	+	+	+	+
Al-2	O	Al	III	+	-	+	-	-	+	-	-	+	+	+	+	+
Al-3	O	Al	III	+	-	+	-	-	-	-	-	-	+	+	+	+
Al-4	O	Al	III	+	-	-	-	-	-	-	-	-	+	+		+
Al-5	O	Al	III	-	-	-	-	-	-	-	-	-	+	+	+	+
Al-6	O	Al	I	-	-	-	-	-	-	-	-	-	+	+	+	+
Al-7	O	Al	II	-	-	-	-	-	-	-	-	-	+	+	+	+

The acronyms and numbers that accompany each strain are just their identification codes; C1-30 isolates - isolates from clinical material; **Al**1-7 - isolates.**A** - Strains from phenotypic research for enterotoxins, from public hospitals in São Paulo; **B** - University Hospital of USP; **C** - Hospital da Clinicas of São Paulo; **D** - Hospital of Santa Maria, Rio Grande do Sul; **E** - Hospital of São Lucas - Rio Grande do Sul; **F** - Hospital da Clinicas of Recife; **G** - Hospital of the Faculty of Medicine of Uberaba; **H** - Laboratòrio mèdico de Santa Luzia, Florianópolis; **I** - Hospital Universitàrio de Belém; **J** - Hospital Universitàrio de Manaus; **L** - Hospital Universitàrio de Vitória; **M** - Hospital Universitàrio de Brasilia; **N**- Strains from a military hospital in São Paulo; **O** - Strains from a food microbiology laboratory in Ribeirão Preto CSF - Cerebrospinal Fluid; LP - Pleural Fluid; Sg - Blood; Al - Foods related to episodes of food-borne infection.

Annex 3: Culture media, solutions and reagents used in PGFE:

Lûria Bertani Agar (Sambrook *et al.*, 1989)

Tryptone	10,0 g
Yeast extract	5,0 g
Sodium Chloride (NaCl)	10,0 g
Bacteriological agar	15,0 g
Distilled H_2O	1000 mL

Tryptone, yeast extract and NaCl were dissolved in 600 mL of distilled water and the pH was adjusted to 7.0 with a 1N NaOH solution. Bacteriological agar was then added and melted to adjust the volume of the solution to one liter with distilled water and sterilized by autoclaving at 121◦ C for 15 minutes.

LB (Lûria Bertani) broth (Sambrook *et al*., 1989)

Tryptone		10,0 g
Yeast extract		5,0 g
Sodium Chloride	(NaCl)	10,0 g
Distilled H_2O		1000 mL

Tryptone, yeast extract and NaCl were dissolved in 600 mL of distilled water and the pH was adjusted to 7.0 with 1N NaOH. The volume of the solution was then adjusted to one liter with distilled water and sterilized by autoclaving at 121◦ C for 15 minutes.

Cont. Annex 3: Culture media, solutions and reagents used in PGFE:

Tampao EC

Tris 6mM	1.5mL Tris 1M pH8.0
1.5mL of 1M Tris pH7.0	
0.1M EDTA	5.0mL of 0.5M EDTA pH8.0
5.0mL of 0.5M EDTA pH7.0	
1M NaCl	29,22g
Brij58 0.5% (polyoxyethylene acetyl ether 20)	2,5g
Sodium deoxycholate0,2%	1.0g
Sarkosyl 0.5% (N-Lauroylsarcosine, sodium salt)	2,5g

Adjust the pH to 7.5, make up to 500mL with deionized distilled water and sterilize by filtration.

Tampa ES

0.4M EDTA, pH9.3	40.0mL EDTA 625mM pH9.3
Sarkosyl 1%	10.0mL of 5% sarkosyl

Prepare in sufficient quantity at the time of use.

CHEF-TE Lid

Tris 0.1M pH7.5	2.5mL Tris 1M pH8.0
2.5mL of 1M Tris pH7.0	
0.1M EDTA pH7.5	5.0mL EDTA 0.5M pH8.0
5.0mL of EDTA 0.5M Ph7.0	

Make up to 50.0mL with deionized distilled water and sterilize at 121°C for 15 minutes.

Culture media, solutions and reagents used in PGFE:

DNS cap

Tris 0.1M pH8.0	5.0mL Tris 1M pH8.0
5mM magnesium chloride	0.250mL of 1M magnesium chloride

Make up to 50mL with sterile deionized distilled water. Prepare at the time of use.

2% low melting point agarose

Low melting point agarose	2g
Deionized distilled water	100mL

Dissolve the agarose in deionized distilled water in a boiling water bath and aliquot into 10mL vials. Sterilize at 121°C for 15 minutes and store at 5°C.

Proteinase K

Dissolve 100mg of proteinase K (GibcoBRL, Life Technologies, USA) in 10mM Tris HCl pH7.5, 20mM calcium chloride and 50% glycerol. Store at -20°C.

Lysostaphin

lisostaphin stock solution

Dissolve 1mg of lisostaphin in 1m l of 20mM sodium acetate buffer pH4.5. Store at -20°C.

Culture media, solutions and reagents used in PGFE:

Lysostaphin use solution

Dilute the stock solution 1:10 in 20mM sodium acetate pH4.5.

Tris-Borate-EDTA (TBE) buffer

TBE 10x

Tris base	121,1g
Boric acid	61,8g
Na2EDTA	3,7g

Make up to 1000mL with deionized distilled water.

Sterilize at 121°C for 15 minutes and store at room temperature.

TBE 0.5χ

TBE 10χ 50mL

Deionized distilled water	950mL

Prepare for use.

Annex 4: Photograph of the gel obtained from pulsed field electrophoresis (PGFE):

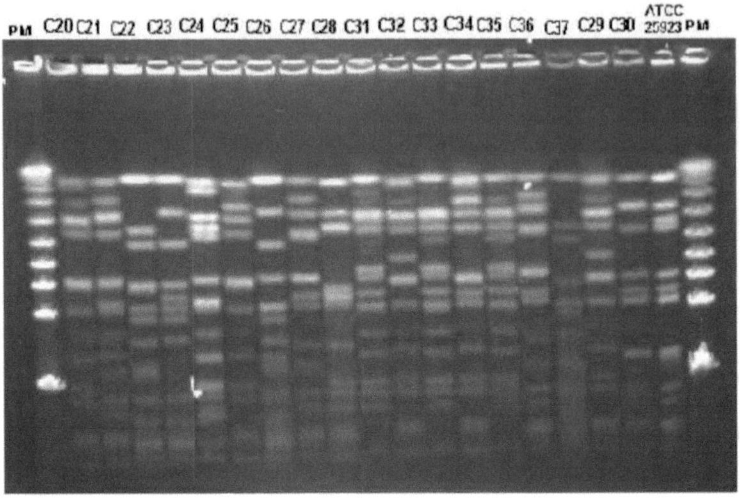

MIX
Papier aus verantwortungsvollen Quellen
Paper from responsible sources
FSC® C105338

Printed by Books on Demand GmbH, Norderstedt / Germany